Arid Environment - Perspectives, Challenges and Management

Edited by Murat Eyvaz,
Ahmed Albahnasawi, Ercan Gürbulak
and Mesut Tekbaş

Published in London, United Kingdom

Arid Environment - Perspectives, Challenges and Management
http://dx.doi.org/10.5772/intechopen.102195
Edited by Murat Eyvaz, Ahmed Albahnasawi, Ercan Gürbulak and Mesut Tekbaş

Contributors
Emilio Ramírez Juidías, Kousik Atta, Saju Adhikary, Manish Kumar Naskar, Bishal Mukherjee, Aditya Pratap Singh, Benukar Biswas, Deborah Rodrigues de Souza Santos, Camila Sarto, Rafael Fernandes dos Santos, Júlia Lôbo Ribeiro Anciotti Gil, Regina Maria Gomes, Evandro Novaes, Carlos Roberto Sette-Junior, Mario Tomazello-Filho, Rafael Tassinari Resende, Matheus Peres Chagas, Carlos de Melo e Silva-Neto, José Norberto Guerra Ramírez, Mohammad Iquebal Hossain, Mohammad Niamu Bari, Kaltoum Belhassan, Murat Eyvaz, Ahmed Albahnasawi

First published in London, United Kingdom, 2023 by IntechOpen
IntechOpen is the global imprint of INTECHOPEN LIMITED, registered in England and Wales, registration number: 11086078, 5 Princes Gate Court, London, SW7 2QJ, United Kingdom

British Library Cataloguing-in-Publication Data
A catalogue record for this book is available from the British Library

Additional hard and PDF copies can be obtained from orders@intechopen.com

Arid Environment - Perspectives, Challenges and Management
Edited by Murat Eyvaz, Ahmed Albahnasawi, Ercan Gürbulak and Mesut Tekbaş
p. cm.
Print ISBN 978-1-80356-551-4
Online ISBN 978-1-80356-552-1
eBook (PDF) ISBN 978-1-80356-553-8

Meet the editors

Dr. Murat Eyvaz is an associate professor in the Environmental Engineering Department at Gebze Technical University. His research interests are applications in water and wastewater treatment facilities, electrochemical treatment processes and filtration systems at the lab and pilot scale, membrane processes (forward osmosis, reverse osmosis, membrane bioreactors), membrane manufacturing methods (polymeric membranes, nanofiber membranes, electrospinning), spectrophotometric analyses (UV, atomic absorption spectrophotometry), and chromatographic analyses (gas chromatography, high-pressure liquid chromatography). He has co-authored many journal articles and conference papers and has taken part in many national projects. He serves as an editor for 50 journals and a reviewer for 120 journals indexed in SCI, SCI-E, and other indexes. He has four patents on wastewater treatment systems.

Dr. Ahmed Albahnasawi is a research fellow in the Environmental Engineering Department at Gebze Technical University, Turkey. Dr. Albahnasawi's graduate work focused on the investigation of the treatability of sequential anoxic-aerobic batch reactors followed by ceramic membrane for textile wastewater treatment; he has published three sci-expanded index journals and participated in three international conferences based on his Ph.D. research. His research interests include the design and application of microbial fuel cells integrated with Fenton oxidation to industrial wastewater treatment/solid waste management and the monitoring of organic micropollutants by both chromatographic and spectrophotometric analyses.

Dr. Ercan Gürbulak is a research associate in the Environmental Engineering Department at Gebze Technical University, Turkey. He received his bachelor's degree in environmental engineering at Marmara University, Turkey in 2005. He completed his graduate work (MSc, 2008 and Ph.D., 2019) at Gebze Technical University. His research interests are the design and application of hydrothermal processes to industrial wastewater treatment/solid waste management and the monitoring of organic micropollutants by both chromatographic and spectrophotometric analyses.

Dr. Mesut Tekbaş is a researcher/lecturer in the Environmental Engineering Department at Gebze Technical University, Turkey. He received his bachelor's degree in environmental engineering at Ondokuz Mayıs University, Turkey in 2003. He completed his graduate work (MSc, 2007 and Ph.D., 2019) at Gebze Technical University. His research interests are the design and application of supercritical water oxidation processes to wastewater treatment/solid waste management and electrochemical analyses.

Contents

Preface

Aridity is the imbalance between long-term average water supply and long-term average water demand. Unlike drought, which is defined as a period of abnormally dry air long enough to cause a serious hydrological imbalance, aridity is permanent, not temporary. A region is arid when it is characterized by a severe lack of usable water inhibiting the growth and development of plant and animal life. Environments exposed to arid climates tend to be devoid of vegetation and are called arid or desert. In the more extreme areas, called extreme arid deserts, the average annual precipitation is below 25mm, under which conditions microorganisms must cope with not only water scarcity but also deadly UV radiation, high and low temperatures, high evaporation rates, prolonged drying times, oligotrophic conditions, and high salinity levels. Arid environments cover more than one-third of the world's land area and represent the most common habitat on Earth after the oceans. Aridity poses a threat to the environment, as well as to the economy and to security, development, food security, and social life around the world. The causes of increased aridity are complex and are thought to be both natural and man-made. Factors such as climate change, population growth, soil erosion, inappropriate irrigation, poor farming methods, soil, water, and groundwater contamination, urbanization, deforestation, improper water management, and desertification of arid and semi-arid zones are among the causes of drought.

A report by the European Commission Joint Research Center on climate change, aridity, and drought states that people around the world feel the effects of climate and environmental crises mostly through water: in the last 50 years, climate and water hazards accounted for half of all disasters and 45 percent of all deaths. The report notes that 90 percent of environment-related casualties during this period occurred in developing countries. Drought is among the greatest threats to sustainable development, especially in developing countries, and a third of the world's population lives in drought-stricken regions. Drought has directly caused the death of approximately 650,000 people in the last half-century. It is reported that approximately 160 million children are currently exposed to severe and prolonged drought, and this situation poses a serious threat to the education, health, and safety of especially women and girls in developing countries. Taking urgent measures on a global scale to combat desertification, water scarcity, loss of agricultural lands, migrations due to desertification and aridity that causes economic losses; determining sustainable land reclamation and ecosystem restoration policies; and steps to be taken against climate change are of vital importance in the fight against aridity and drought.

This edited collection contains seven chapters. Chapter 1 is a general introduction to the topic of aridity and relevant definitions. Chapter 2 highlights drought and water stress in Northern African countries and outlines the main approaches to improving water management in every country of the region. Chapter 3 focuses on the initiatives taken by the Barind Multipurpose Development Authority to transform the arid-like Barind area into a green arable landscape through efficient water management.

Chapter 4 reviews remote sensing tools in agricultural applications. A novel model predicting transported soil volume through the presence of water which is presented in Chapter 5 will be a point of reference in future studies involving rocky planets. Chapter 6 illustrates the impacts of the industrial and technological revolution on territories and cities in arid environments with a case study of the Antofagasta region in the Atacama Desert, Chile. The final chapter evaluates the wood quality of five elite four-year-old Eucalyptus spp. clones from a clonal test installed in a pulp-producing region of seasonal drought stress in central-western Brazil.

Murat Eyvaz
Associate Professor,
Department of Environmental Engineering,
Gebze Technical University,
Kocaeli, Turkey

Dr. Ahmed Albahnasawi, Dr. Ercan Gürbulak and Dr. Mesut Tekbaş
Department of Environmental Engineering,
Gebze Technical University,
Kocaeli, Turkey

Chapter 1

Introductory Chapter: Arid Environment

Murat Eyvaz and Ahmed Albahnasawi

1. Introduction

Arid environments cover more than one-third of the world's land area and represent the most common habitat on Earth after the oceans. Aridity poses a threat to the environment, as well as the economy, security, development, food security, and social life around the world. The causes of increased aridity are complex and are thought to be both natural and man-made. Factors such as climate change, population growth, soil erosion, inappropriate irrigation, wrong farming, soil, water, and groundwater contamination, urbanization, deforestation, improper water management, desertification of arid and semiarid zones appear as causes of drought.

This book covers a wide range of scientific research studies, from water management to groundwater management, from land rehabilitation to soil reclamation, which will help prevent and minimize man-made aridity. In addition, many studies related to aridities such as environmental education, environmental awareness, sustainable development, and management policies and plans are also welcome.

2. Aridity

Aridity is the imbalance between the long-term average water supply and the long-term average water demand [1]. Unlike drought, which is defined as a period of abnormally dry air long enough to cause a serious hydrological imbalance, aridity is permanent, not temporary. Aridity is a condition in which the amount of usable water in an area is reduced to such an extent that it hinders or prevents the growth and development of plants and animals. Regions with arid climates lose or tend to lose their vegetation. These regions, which are generally located close to the equator, are called xeric, arid, or desert according to their aridity levels. In the more extreme areas, called extreme arid deserts, the average annual precipitation is below 25 mm, under which conditions microorganisms must cope with not only by water scarcity but also by deadly UV radiation, high and low temperatures, high evaporation rates, prolonged drying times, oligotrophic conditions, and high salinity levels. Aridity is often evaluated with the aridity index (**Table 1**) and aids in determining whether there is a water shortage in the region and deciding the measures to be taken in case of a possible arid climate [3].

	Climate type	Aridity i
Dry land subtypes	Hyper-arid	AI < 0.05
	Arid	0.05 ≤ AI < 0.2
	Semiarid	0.2 ≤ AI < 0.5
	Dry subhumid	0.5 ≤ AI < 0.65
Non-dry lands	Humid	AI ≥ 0.65
	Cold	PET* < 400 mm

**PET: Potential evapotranspiration*
Adopted by [2]

Table 1.
Climate classification and dry land subtypes based on the Aridity Index.

3. Sustainable development of arid regions

The main causes of dry weather in arid ecosystems are high regional temperature caused by the sun's rays, high pressure caused by pressure centers, ocean coast streams, high mountain ranges and high plateaus, lack of exposure to marine influences, absence of upward movement of air, and air turbulence [4]. In addition to the adverse effects of the climatic factors, arid areas also have some potential if they are well analyzed and defined. Among them are solar and wind energies; salt, sand, and gravel pits; coal, uranium, copper, and zinc mines; natural touristic sights and horticultural activities. To evaluate these potentials and to realize possible industrial and tourism investments, sustainable development plans and activities in arid areas should be investigated [5]. Environmental management and sustainable development approaches in arid ecosystems are summarized in **Figure 1**.

4. Combating climate change and drought

Climate change is one of the most important global challenges of our time, with its borderless nature that affects all countries regardless of their level of development. In the report titled "Human Cost of Climate-Related Natural Disasters" published by the UN in October 2015, 90% of a total of 6,457 natural disasters recorded in 20 years are caused by floods, storms, heat waves, droughts, and other extreme climate movements. It is stated that 606 thousand of people have lost their lives and 4.1 billion people have been affected since 1995 due to disasters caused by extreme climate movements. In total, 80% of the poorest group, which receives the least share of income distribution worldwide, lives in rural areas doing small-scale agriculture and animal husbandry. The decrease in precipitation due to climate change and the pressure on natural resources caused by soil degradation can force these vulnerable groups to migrate, which triggers instability and brings security risks [6]).

Aridity/drought and desertification are important environmental challenges of our time and are directly related to climate change. Desertification, drought, and land degradation because of extreme weather conditions triggered by climate change are a global test that threatens the living space and most basic livelihood of most of the world's population and may pose a food security risk, rather than an environmental

Figure 1.
Environmental management and sustainable development approaches in arid ecosystems (Adopted from [4].

problem. Balancing land degradation is key to achieving the Sustainable Development Goals (SDGs). On the other hand, sustainable land management is also of great importance for combating climate change, because the soil whose organic component is preserved acts as a sink with the ocean and forests by trapping the carbon in the atmosphere. It is possible to realize approximately one-third of the greenhouse gas reduction potential with land degradation balancing and sustainable land use. The most important tool for balancing land degradation and combating desertification is the "United Nations Convention to Combat Desertification in Countries Exposed to Severe Drought and/or Desertification, Especially in Africa" (UN Convention to Combat Desertification – UNCCD) adopted in 1994. The Convention is the only binding international agreement linking the issue of environment and development with sustainable land management [6].

5. Conclusions

Within the scope of combating climate change, the transition to a low-carbon economy on a global level envisages a radical transformation that will change people's lifestyles, production, and manufacturing methods. For this reason, efforts to combat and adapt to climate change should not be perceived as a mere environmental problem. This struggle can directly affect the growth strategies, energy policies, health and agriculture-related programs, use of water resources, food security, transition to a low-carbon economy, and sustainable development goals to be followed by developed and developing countries and may be decisive in their development. To achieve these goals, developed countries need to fulfill their commitments to financing, technology transfer, and capacity building. To combat aridity and desertification, it is necessary to determine the situation on a global scale and to force countries to work on combating drought and desertification. In addition, national and global funds should be mobilized to promote good practices, contribute to sustainable development in countries affected by drought and desertification, develop cooperation in combating drought and desertification, and support studies in this field.

Author details

Murat Eyvaz* and Ahmed Albahnasawi
Department of Environmental Engineering, Gebze Technical University, Kocaeli, Turkey

*Address all correspondence to: meyvaz@gtu.edu.tr

References

[1] Maliva R, Missimer T. Arid lands water evaluation and management. 1st ed. In: Förstner U, Rulkens WH, Salomons W, editors. In the Series of Environmental Science. Heidelberg: Springer Berlin; 2012. pp. 21-37

[2] Middleton NJ, Thomas DSG. World Atlas of Desertification. 2nd ed. London: Arnold; 1997

[3] Paparrizos S, Maris F, Matzarakis A. Integrated analysis and mapping of aridity over greek areas with different climate conditions. Global Nest Journal. 2016;**18**:131-145

[4] Estelaji A. Approaches to sustainable development of arid regions. Desert. 2008;**13**:53-58

[5] Estelaji AR, Shariat Panahi MV. A model for environmental management in arid areas, with focus on tourism development: A case study of desert areas in Iran. WSEAS Transactions on Environment and Development. 2008;**4**:1089-1098

[6] Republic of Türkiye Ministry of Foreign Affairs. 2022. Available from: https://www.mfa.gov.tr/default.en.mfa [Accessed October 15, 2022]

Chapter 2

Managing Drought and Water Stress in Northern Africa

Kaltoum Belhassan

Abstract

Northern Africa is a region surrounding the northern portion of the African continent. Northern Africa consists of five countries Egypt, Algeria, Morocco, Tunisia, and Libya. Northern Africa has semi-arid and arid desert climates and low rainfall. Over the past four decades, many areas in Northern Africa have faced drought which has become more widespread, prolonged and frequent due to climate variability and which may expedite a shortage of water and to a decrease in the land areas suitable for agriculture. In fact, limited water reserves, growing population and droughts are the main factors reflected in the increased consumption of freshwater. It is critical to understand a balance between water demand and supply by managing drought and water stress in the region.

Keywords: Northern Africa, drought, water stress, managing

1. Introduction

Earth is referred to as the water planet because 75% of Earth's surface is covered by water. More than 97 per cent of the Earth's water is contained within the oceans as saltwater. Less than 3 per cent of this water is freshwater, and most of it exists in glaciers, underground, lakes, rivers, and swamps. Water means life. Water is the key to sustainable development and is crucial for socio-economic and for human survival itself. The rise in population will increase water consumption which will intensify water demand, and this will limit the amount of water available per person (drinking, irrigation, industries and municipal needs). The most water-scarce areas are those with lack of freshwater resources and higher population growth rates. Water is also the core of climate change adaptation, serving as the major link between society and the environment. Climate change is already having many effects on the Earth and particularly on the water's Earth. In future decades, climate change is predicted to affect availability and distribution of water (precipitation, river, groundwater…). Thus, the handling of water will become ever more critical. It is further estimated that by 2025, more than half of the world's population will be living in water-stressed areas due to the world population growth, which is predicted to reach ~9.7B by 2050, causing further stress on water globally. Although water stress is a universal phenomenon, this chapter focuses on the continent of Africa with approximately 1.37B inhabitants [1] and more specifically the region of Northern Africa which includes a total of five countries, viz. Egypt, Algeria, Morocco, Tunisia and Libya (**Figures 1** and **2**). Despite, each country in Northern Africa with its geological characteristics (geographic and climatology) and water management history, Algeria, Morocco and Tunisia have many common characteristics [4]. A map of Northern Africa with water stress is shown in **Figures 1** and **2**.

Figure 1.
Map of Northern Africa [2].

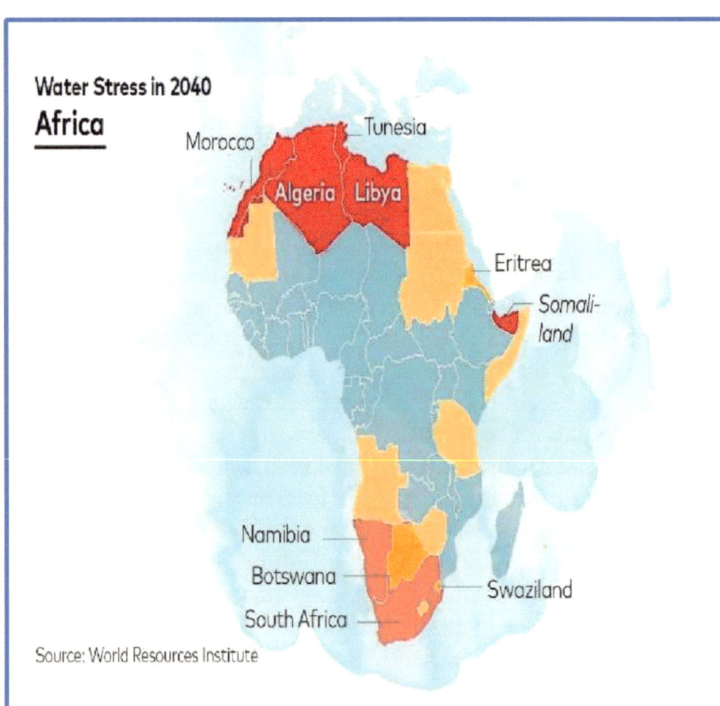

Figure 2.
As comparison projected water stress in Africa in 2040, as per Water Resource Institute [3].

In Northern Africa, humans' activities depend essentially on water availability. Climate change impacts Northern Africa's water in different ways. It is a principal contributor to droughts (low precipitation and rising in temperature). Droughts are complicated phenomena generally associated with greatly reduced precipitation. Droughts are recurrent and often devastating threatening people's livelihoods in Northern African countries. Droughts will likely continue to decrease the average water availability in an already water-scarce region. These poor water management threats have led North African countries to overdraw water from rivers and aquifers and thus degrading already scarce water resources. Climatic variability in Northern Africa (rise of temperatures and low rainfall) will lead to approximately 22% of lack of sufficient water by 2050 [5], which will increase more the long periods of droughts in this region of the world [6].

The main objective of this chapter is to highlight drought and water stress in Northern African countries along with outlining the main approaches and methods applying to better water managing in every country of Northern Africa.

2. Population growth

Population growth is increasing number of people over time for a particular place, depending on the balance of births and deaths. Population in Northern Africa has grown from 55.65 million to 205.11 million people, the North Africa's population is more than tripled in size between 1960 and 2021. Egypt is the highest population, with over 104 million people. In 2021, the total population of Libya was about 6,96 million inhabitants. It is the country in Northern Africa which has the lowest population (**Table 1**).

In Northern Africa's region, population growth increases because the fertility rate exceeds the death rate and thus water consumption is increasing inexorably through a combination of different water uses (drinking purposes, irrigation and other humans uses) and will further limit the freshwater availability per person. The most water scarce areas in Northern Africa are typically those, with limited source of freshwater and highest population density (the number of individuals per square mile or in square kilometers of land area), where people are currently unable to meet their basic water demand.

3. Drought and water stress in Northern Africa

Northern African region is characterized by considerable demographic growth. Northern Africa has hot semi-arid and arid desert climates and low rainwater. This region is predicted to face rising temperatures and decreasing rainfall. The recurrent droughts whose frequency has increased over the past 40 years can lead to water stress situations. While the overexploitation of available natural water reserves can exacerbate the consequences of droughts and thus leading to high depletion in its groundwater reserves.

3.1 Egypt

Egypt's climate is extremely dry and receives a very low amount of rainwater because Egypt is predominantly desert [7]. Water resources in Egypt are limited and

Regions	1960	2021
Egypt	26.63 million	104.26 million
Algeria	11.06 million	44.62 million
Morocco	12.33 million	37.34 million
Tunisia	4.18 million	11.93 million
Libya	1.45 million	6.96 million
Northern Africa	55.65 million	205.11 million

Table 1.
Northern Africa's population 1960–2021 [1].

around 98% of Egypt's water reserves originate outside of its borders. Main Egypt's freshwater reserves are the Nile River which provides the region with around 93% of its water demand, precipitation, and ground water [8]. Climate change and its effects (droughts) in Egypt are predicted to make more stress on water reserves.

Nile River - Because the rainwater in Egypt is very scarce, more than 95% of Egyptians live along the Nile or in its teeming delta for all water needs. However, the Nile River's land is threatened by climate change vulnerability which is expected to rise the frequency and intensity of floods and drought incidences in Egypt. Actually, the Nile River is predicted to decline by 40–60% of its current flow. Therefore, agricultural areas in Egypt have been subject to extensive and increasing water scarcity [9].

Rainwater – Egypt's agricultural land is concentrated in a narrow strip along the coast where more amount of rainfall, and continuously declines southward [10]. Egypt has a mild winter season in the form of scattered showers along coastal areas. The mean annual rainfall is between 0 mm in the desert to 200 mm in the north coastal region with an annual average of about 12 mm. The total amount of rainwater may reach 1.8 billion m^3 per year [11]. It is predicted that the mean annual precipitation will decline not only in Egypt but also in almost Mediterranean African countries. Furthermore, changes in precipitation patterns combined with sea level rise may further lead to decline in agriculture production and cause more stress on Nile River.

Groundwater is a vital resource especially in desert areas, besides providing drinking water groundwater is used for irrigation to grow crops and also has several other purposes. Groundwater ranks as the second source of natural water resources, after the Nile and constitutes around 12% of water supplies [12]. Egypt's groundwater aquifers are considered a non-renewable resource and they are of variable importance for exploitation. The increasing demand for Egypt's water reserves has placed groundwater resources under widespread pressure subjecting it to several issues such as depletion of its water quantity and degradation of its water quality. Different human activities such as the seepage of rainwater, irrigation and drainage and other effluents commonly affect not only the quantity of groundwater resources but also its chemical water quality [13]. On the North Coast of Egypt, there is a threat of saltwater intrusion into coastal aquifers which is caused by water withdrawals from coastal aquifers and up-coming near coastal discharge/pumping wells. On the other hand, Egypt faces several serious risks from climate change that impacts almost Egypt's groundwater recharge rates and thus affects the availability of fresh groundwater (decrease in amount of precipitation and overexploitation of aquifers).

3.2 Algeria

Algeria is the biggest country in Northern Africa. The southern region forms about 80 per cent of Algeria's land which is almost entirely the Sahara Desert with an arid climate. However, the northern part of Algeria's territory is less arid. More than 34% of Algerian people are living in rural areas and most of them are concentrated in coastal zone [14]. During the past century, Algeria has been subject to frequent periods of drought which has clearly led to the degradation of the water reserve (quantitative and qualitative) in northwestern and central plains. Thus, these droughts have an adverse impact on meeting the water needs of all socio-economic sectors especially agriculture; posing a significant risk to farms and leading to yield reduction.

Rainwater - Temperatures in Algeria rise during the last decades and can become more acute than expected, with an important decrease in annual rainfall, sometimes various persistent droughts. Reduced rainfall is about 18 to 27% and the dry season has increased by two months during the last century [15].

Groundwater - The rainfall deficit has resulted in a decrease in the water volumes stored in dams. This situation of insufficient freshwater to satisfy requirements and droughts events in Northern Algeria continue to cause significant risks and widespread pressure on most aquifers; resulting in groundwater-level decline which may reach around 20 m or more in certain aquifers [16].

3.3 Morocco

Climate is varying considerably across Morocco's northern to southern areas. The northern coast-central areas have Mediterranean climates characterized by hot, dry summers and cool, wet winters. At high-altitude locations, the climate is humid and temperate. In the south and west parts, the climate is Saharan. Droughts are becoming more frequent and severe in Morocco and impact negatively water resources and agricultural production [17]. Mean annual temperature has increased by 0.8°C (1960–2005) [18].

Rainwater - Morocco is commonly characterized by extremely high spatial–temporal rainfall variability. The northwest part of Morocco receives more precipitation than others. The mean annual precipitation varies considerably; the relief areas can receive approximately 800 mm of rain per year. Nevertheless, the adjacent plains areas can extremely receive 300 mm of rainwater per year [19]. Due to climate variability observed over the past decades in Morocco (1960–2005), Mean annual rainfall has decreased with an amount ranging between 3% and 30%, with a drop of 26% in the north-western region of the country [18]. In fact, Morocco suffers from its worst drought over the past 40 years with low precipitation resulting in at least ~50 per cent overexploitation of aquifers.

Groundwater stretches more than 80,000 square kilometers and represents around 20% of Morocco's water reserves [20]. Morocco has ninety-six aquifers, twenty-one of which are deep and seventy-five shallows. Morocco faces its worst drought in last four decades resulting in (1) decline in rainfall, (2) decline in River flows which reached approximately 20% within the periods 1970–2006 and 1950–2006 and more than 70% in certain parts of Morocco and also (3) depletion of groundwater aquifers [21] ranging between 0.5 and 2 m per year (low groundwater recharge and over-expansion of agricultural activities) [22]. The Mikkes basin is an example of Moroccan basin. It is situated in the North-central of Morocco and had a high depletion in its water reserves: a rainfall deficit reached 76% (1970–1979 and 1980–2000); this high rainfall deficit had led to overexploitation of Mikkes aquifers and thereby springs and River flows depletion [23–27]. Furthermore, Morocco's aquifers groundwater quality assessment shows a deterioration, as a result of combined pressures of climate change (seawater intrusion) and human activities (nitrate pollution).

3.4 Tunisia

Tunisia is situated between Algeria (on the west and southwest) and Libya on the southeast (**Figure 1**). Tunisia's climate varies with location: The north region is with a sub-humid to semi-arid climate. The central part is with semi-arid to arid climate and the southern part has a desert climate. Tunisia is vulnerable to climate change resulting rise in temperature, a reduction in rainwater and sea level rise. Over the past decades, temperature increased by approximately 0.4°C per decade. In southern Tunisia, droughts become increasingly more intense and frequent, while Tunisia's Mediterranean coast has been treated with rising sea level and flooding [28]. Between 1900 and 2000, the mean annual temperature increased by around 1.4°C [29].

Rainwater - As Tunisia is bound on the northeast by the Mediterranean Sea, on the south - the southwest by the Sahara (**Figure 1**), the precipitation is variable from the north to south and from east to west. Northern Tunisia is the rainiest part as mean annual rainfall reaches more than 400 mm in the extreme north and 1500 mm in the extreme northwest. However, the centre of Tunisia has average precipitation ranging between 150 and 300 mm [30]. The Tunisian southeast is characterized by an arid Mediterranean climate, with average precipitation varying between 100 and 200 mm per year [31]. Actually, precipitation amounts have changed in Tunisia during the 20th century and the region has experienced several severe droughts. Since the fifties, the annual totals of precipitation have declined by 0.5% per year in northern Tunisia [32].

Groundwater - As Tunisia has been suffering from increased recurrent and frequent droughts which led to high rainwater decline. Many farms rely on wells (groundwater) to fulfil their water needs. This will most likely extend to the over-use of groundwater resources where many aquifers are already experienced; putting more pressure on groundwater means that the rate of pumping is greater than the rate of infiltration. Consequently, the level of certain aquifers drops so much and thus many wells run dry. Actually, by 2030, the overexploitation of Tunisian coastal aquifer (due to agricultural activities) will drop from 28% [33] and 50% of coastal aquifers and will be salinized due to sea-level rise [34]. Therefore, this country needs urgent and practices approaches for better water management to avoid more water reserve deteriorations.

3.5 Libya

Libya is a country located in North Africa and it is the fourth-largest one in all African continents (**Figure 2**). About 95% of Libya consists of desert (the Sahara) and under 2% of the land is arable. Libya has two distinct climates: One in the south is characterized by hot arid Sahara climate and the other is influenced by the Mediterranean Sea resulting moderated climate. Climate change has many effects including rising temperatures and more severe frequent droughts and floods. The limited natural resources (water and soils), the desert lands and the drought events are all the main drivers which will force Libyan farmers to abandon their farms amidst water stress and therefore the yields of rainfed agriculture will be severely low.

Rainwater - Libya is one of the driest regions on Earth with no permanent rivers flowing through its boundaries. About 96% of Libyan land surface receives annual precipitation which cannot reach an amount of 100 mm. Less than 2% of the Libya region receives enough rain to support agriculture, and only a narrow ribbon along the coast receives more than 100 mm of rain per year. Furthermore, the effects of global climate change in Libya include more rising in temperatures, a drop in precipitation amount which is already very low and prolonged period of droughts which will produce increased aridity. For the 66 years (1945–2010), temperature data in Libya showing that the mean annual temperature has risen [35].

Groundwater freshwater reserves in Libya originate from 4 aquifers: Kufra, Sirt, Morzuk and Hamada that provides over 90% of Libya's water. These aquifers are likely to be even more important as drought increases and rainfall decreases. Increase in drought tendency is a principal factor in lack of rainwater (or water availability). This water scarcity is expected to cause more aquifers depletion (quantitative and qualitative) in the region. The intensive extraction of groundwater (GW) in coastal aquifers is causing a reduction in the availability of freshwater outflow to the sea and creates local water table depression, causing saltwater intrusion [36], resulting in deterioration in groundwater quality.

4. Agriculture

Northern Africa is reported to be among the world's most water-scarce regions where drought is a principal climatic factor (rising temperature and drop in rainwater) reducing agricultural production [37]. Agriculture is the most vulnerable economic activity. It is considered as a major challenge in Northern African countries, particularly for farmers whose livelihoods rely on rainfed farming. Farmers in Northern Africa have to increase crop production to meet the rising needs for food.

Egypt is predominantly desert, with only 4–5% of land used for the Egyptian people to live and produce food [38]. Egypt is a water-scarce country (with less than 1000 m^3 of freshwater per year) and it is near the edges of absolute water scarcity (less than 500 m^3/year). Agriculture plays a significant role in the Egyptian economy as it is the sector which plays a crucial role in food's productivity. This sector provides livelihoods for about 55% of Egypt's population, which is largely rural [39]. Agriculture consumes between 80 and 85% of water resources. More than 90% of River Nile water goes towards agricultural productivity. As Egypt's population still grows, the economy is expanded and significant drought severity, all these drivers and others are contributing together to more growth in freshwater demand and render the whole of the Nile valley vulnerable. Since water availability (quantitative and qualitative) is the major factor of agricultural production, the Egypt authority needs a sustainable water strategy to better ways of resolving the shortages of water and is through good agricultural water management.

Algeria is a poor country in water resources because of the irregularity (insufficient and unequally distributed) of water supplies. The annual water supplies drop below 1000 m^3 per person. Agriculture is the largest using sector of water and is increasingly subject to water risks. Almost 25% of the Algeria population is engaged in agriculture sector. Water demand is predicted to further increase because of population pressures increase, intensive agriculture, economic growth and high drought risk under climate change. Thus, developing a mix of strategies that increase water supplies, manage water demands, and reduce long-term pressure on water is urgent more than ever before.

Morocco - Agriculture represents almost 15% of Morocco's GDP [17]. Agriculture is the primary user of water, accounting for 80 percent of withdrawals. Morocco suffered severe water shortages under its worst drought in 4 last decades and which is prompted by expansion in water needs (decrease in rainfall and bad water management). Morocco expects to reach absolute water scarcity (the annual water supply drops below 500 m^3 per person) and people are expected to live under extreme water stress in less than 25 years [19]. Actually, various Innovative irrigation practices can help in reduction of water uses in agriculture. Nevertheless, they are expensive for small farmers to afford.

Tunisia is one of Northern African countries suffering from water stress as it has limited surface water reserves, low precipitation and thus a great increase in agriculture's dependence on groundwater withdrawal. As Tunisia's population increased from 4.18 million (1960) to 11.93 million (2021) (**Table 1**) and also drought frequency so all these factors and others have drastically increased food needs to be grown, which requires more water. About 80% of Tunisia's freshwater resources are used for agricultural exploitation; Over 76% of groundwater is used by agricultural sector. However, less than 24% of irrigation water use come from surface water [40]. Tunisia witnessed its worst drought in 50 years from 1999 to 2002, which caused more increase in water consumption and affected agricultural producers [41]; resulting in deterioration of the quality and quantity of groundwater reserves.

Libya is the second-largest country in Northern Africa, and it is almost entirely covered by the Libyan Desert (**Figure 2**). Thus, most residents of Libya live in the coastal regions where more chance of water availability. Groundwater constitutes the major freshwater supply in Libya [42] and it constitutes around 98% of the total freshwater demand [43]. Irrigated agriculture is also the largest using sector of freshwater; it consumes more than 80 percent of freshwater. The shortage of clean and freshwaters in Libya have led to an overall situation of overuse of groundwater. During the last 4 decades, the irrigated land has increased considerably which is causing more pressure on water availability which is already very stressed.

5. Managing drought and water stress in Northern Africa

Water stress is a growing problem worldwide. In Northern Africa, the demand for water is likely to increase while water supplies are reduced. This is driven by a combination of the rising population (coupled with economic growth) and more frequent periods of droughts which the region recognized especially over the last 4 decades, rendering water availability in the future uncertain. Groundwater tends to be overexploited and polluted to meet growing water consumption. Water stress and drought situations around Northern African countries constitute widespread two major phenomena (natural and human-made) that challenge water security, hence, all need to work together for better solutions of water issues as such, i.e., reuse of drainage wastewater, water desalination and rainwater Harvesting (RWH).

5.1 Reuse of drainage wastewater

Rising population and long period of droughts exerted high pressures on renewable freshwater resources in Northern Africa over the last decades. This will lead to markedly greater competition between urban areas and farmland for water and thus producing more wastewater. Actually, the reuse of drainage wastewater is a great way to help protect the environment by reducing dependency on freshwater and meeting rising water consumption. Wastewater treatment is a process used to remove contaminants, micro-organisms and other types of pollutants from wastewater or raw sewage and convert it into an effluent that can be returned to the water cycle. Reusing wastewater is an obvious solution for the future to reduce water shortages in several regions on the Earth such as Northern Africa. After treating sewage, the treated wastewater can be reused for several applications such as irrigation purposes.

Egypt – Egypt's freshwater resources are limited. The gap between freshwater demand and supply is intensified by various combined factors including rising population and drought frequency. Reuse of sewage effluent in agriculture represents an opportunity that can alleviate the water stress on limited natural water reserves in Egypt. This process started in 1920 [44] and it is relatively less-infrastructure requirements to be constructed and cheaper option. The 1975 water policy for reuse of wastewater discharge had a target to decrease water stress on the Nile system and hence to increase cultivated areas and meet Egypt's growing food. Currently, this system is broadly applied in Delta region and produces approximately 4.0 BCM/year of wastewater discharge to be mixed with the freshwater of main canals. The Government's Planning for the Future is to reuse an additional 3 BCM/year. Besides these benefits, it deserves consideration of some disadvantages including the use of untreated wastewater for crop irrigation can also cause soil hardening and shallow groundwater contamination [45]. Therefore, Egypt started using treated domestic sewage wastewater treatment systems which will increase water recovery from wastewater and meet environmental regulatory requirements and thus protect freshwater bodies and biodiversity. SUEZ is providing Egypt with different types of sewage treatment plants including that of Gabal El As-far, on the eastern bank of the Nile [46].

Algeria - Over the past 4 decades, Algeria has experienced water stress which becomes acute due principally to high population growth rates, long periods of droughts and bad water management. The reuse of treated sewage effluent represents a valuable solution to conserve natural resources and reduce the consumption of freshwater, especially in the agricultural sector. In 2005, Algeria started using treated wastewater as alternative resources that are able to satisfy the needs of water demand in agricultural sector and promote the coordinated development of integrated water management systems [47]. At present, Algeria could indirectly improve water supply and increase water availability by reusing around 484 Hm^3 of wastewater, among which only 425 Hm^3 are subjected to water treatment (removes contaminants and undesirable components). The rest simply undergoes dilution in the natural environment [48].

Morocco is highly susceptible to prolonged shortages in the water supply (droughts). To reduce the impact of drought and population growth on water consumption, the Morocco government has adopted a series of legislative measures and institutional reforms to better Integrated Water Resources Management (IWRM). Since 1960s, the country has contributed significantly to the mobilization of its hydraulic capacities. Applying treated sewage effluents (TSE) in agricultural irrigation in regions suffering from water scarcity like Morocco is a non-conventional water reserve to alleviate water stress and help in saving freshwater for drinking and for improving crop productivity. Thus, a necessity for a better water resources economy, only the rate of about 12% of treated wastewater is currently recycled but this rate reached 22% in 2020 and may achieve around 100% by 2030 [49].

Tunisia is suffering from high water stress due to many contributing drivers including the region has a Mediterranean arid climate. Tunisia's water reserves are limited. Tunisia is highly vulnerable to the adverse effects of climate change (increase in temperatures and aridity with decreasing rainwater). Actually, Tunisia is determined to promote wastewater reuse and satisfy its water demand for agricultural sector and other uses and this through many ways including improving the status of existing water resources, reducing the effluent of wastewater treatment plants to the sea and raising awareness. The government policy strongly supports sewage treatment plants and incentivizes wastewater reuse. In 2009, around 63 Mm^3 have been reused directly for irrigation and the total agricultural area equipped with sewage treatment plants was around 8065 hectares [50].

Libya is facing extremely high baseline water stress. With the rising population and demands for freshwater, reusing wastewater is an increasingly sustainable and acceptable practice to satisfy water needs. Thus, Libya government is increasing efforts to enhance wastewater treatment plans in order to cope with water scarcity in the region and generate sufficient water, especially in irrigation. There are about 23 wastewater treatment plants in Libya but only 10 of them are working and in operation [51].

5.2 Water desalination

Seawater desalination is the process of changing seawater or brackish into usable water or pure water by which the dissolved mineral salts in water are removed. Northern African countries are suffering from water shortages for agricultural purposes and other uses. So, there are alternative potential water supply sources to meet growing water needs. Desalination is among the most sustainable alternative and extreme solutions that can solve water stress issues in Northern African countries, although it is an energy-intensive process that can be very expensive.

Egypt has the following coastlines: Northern coastal border is on the Mediterranean Sea and east coast border is on the Red Sea. This country is using seawater desalination as a major and sustainable source of water supplies and development (abundance of energy). However, this practice has been given low priority because it is affected significantly by many different factors including water quality, technical application and methods, energy-consuming, plant capacity and plant availability [52]. The capacity of desalinated water in Egypt is approximately 0.03 BCM/year [53]. Additionally, the Egypt authority has involved both sectors (the public sector and the private one) to work together to a better agricultural water resource management through applying modern technologies for desalination such as distillation, reverse osmosis (RO) and electrodialysis.

Algeria - Desalination plants provide water that can be safe to use in irrigation. Algeria is using desalination as a viable resource during the intense time of drought. In fact, it is a great practice in Algeria as it can relieve water stress for irrigation purposes and also for other water uses for daily processes. Reverse osmosis technology is the most convenient and effective filtration method used for desalination that represents approximately around 95% [54].

Morocco - Water desalination is practically a solution for Southern Moroccan; most part of inhabitants suffers from shortages of potable water and inadequate precipitation. The following of some Southern Moroccan cities which adopt these water desalination solutions: Boujdour has used Multi-Effect Distillation and also Mechanical Vapor Compression solutions to provide a total capacity of around 250 m^3/d and Boujdour has used Sea Water Reverse Osmosis solutions to provide a total capacity of around 800 m^3/d. Laayoune has used Sea Water Reverse Osmosis solutions to provide a total capacity of 7000 m^3/d [55].

Tunisia and especially the centre and south parts of it are suffering from water scarcity and droughts periods as the low rainfall. Water desalination for irrigation seems to be a promising solution to fulfill the increased demand for freshwater. The four major desalination plants have been inaugurated to help are Kerkennah (1983) with a total desalination capacity of 3300 m^3/day; Gabes (1995) with 22,510 m^3/day; and 2 stations in Jerba-Zarzis (1999) with 12,000 m^3/day. Furthermore, there are sixties smaller plants used to help in providing water for industry uses [56].

Libya faces severe water stress problems caused mainly by the limited freshwater bodies and also drought. Water desalination is a particularly advantageous alternative freshwater source that can reduce water stress in Libya. Since the sixties, Libya has been using desalination because of its an increasingly viable alternative as it is regarded as an extreme solution for water supply. In addition, this desalination technology is widely implemented to produce freshwater over the past decades to meet the increasing demands of water for irrigation purposes and other uses. Currently, Libya has 21 operating desalination plants with a capacity of 525.680 m^3/d [57].

5.3 Rainwater Harvesting

Many people in different areas of the world such as Northern Africa suffer from lack of access to safe and clean drinking water. To have access to safe potable water, huge investment costs and expenditures are needed. Roof-water or RWH is a method of collecting and conserving rainfall for future usage. The harvested water can be stored, utilized in various ways or directly used for groundwater recharge. Roof-water is an old method that has been adopted in different regions on the Earth and especially Northern African region [58]. RWH is a viable solution to help meet the growing demand for water. It can improve water productivity by collecting rainwater from impermeable surfaces (rooftops) and storing it in containers (tanks or cellars) for future uses. Additionally, RWH helps in reducing floods and soil erosion and may reduce agricultural drought risk.

Egypt – In Egypt, natural water reserves are limited. Actually, Egypt is under water stress, a problem that can be partially alleviated to meet people's needs by using RWH as good alternative to non-conventional water resources. High potential of RWH was built in Northern Egypt including Alexandria city which can help in reducing water usage home and can satisfy approximately 12 percent of its future supplementary domestic purpose. Nevertheless, in the central – south part of Egypt, the precipitation is irrelevant to be harvested [59].

Algeria - Water is naturally scarce in Algeria because of the following factors: very low rainwater, human growth considerations, droughts and thus water demand is continuously increasing. Using rainwater harvested in Algeria from houses roofs appears a great sustainable promising solution to lack of water and droughts to satisfy water needs in areas where rainfall is uneven or unequal, such as in Souk Ahras city.

Morocco has adopted fog water harvesting system-based NGO Dar Si Hmad as the most promising alternative system for sustainable freshwater resources management to minimize the shortages of water. This system is the world's largest operational fog-water harvesting system. It delivers a good solution for hundreds of rural residents who are suffering from water shortages to satisfy their basic water needs [60].

Tunisia - RWH is a solution to help reduce freshwater consumption by utilizing – rainwater from the roof. It is a satisfactory alternative practice in southeast Tunisia which suffers much from water shortage (mean annual precipitation values from 100 to 200 mm). The Tunisian authority promotes many water harvesting techniques including surface runoff water harvesting, floodwater harvesting and spreading irrigation [4].

Libya has consistently suffered from water stress and droughts. RWH is one of the major options to provide more water. RWH is a sustainable way in Libya's coastal part that can deal with shortages of water and meet the increasing water demand (clean water).

6. Conclusion

Northern Africa is one of the world's most water-scarce regions. The freshwater resources in Northern Africa region are limited. Due to many factors including population growth, recurrent and frequent droughts over the last 4 decades and poor freshwater management, water demands will increase to satisfy different human uses such as irrigation purposes. Thus, water scarcity in the region is going to increase in future years and be more worsen. Hence the need to embrace the best water managing demand and through several unconventional water resources such as reuse of drainage wastewater, seawater desalination and rainwater harvesting. Indeed, managing water scarcity in Northern Africa in a sustainable, efficient, and equitable ways is paramount to tackling Northern Africa's water stress.

Author details

Kaltoum Belhassan
Independent Researcher, Dewsbury, UK

*Address all correspondence to: kbelhassan@yahoo.co.uk

References

[1] UN, United Nations, Department of Economic and Social Affairs, Population Division. World Population Prospects 2019, Online Edition. Rev. 1. 2019. Available from: https://worldpopulationreview.com/continents/africa-population

[2] MSA. Maghreb Studies Association. The Word MAGHREB. 2004. Available from: http://www.maghreb-studies-association.co.uk/en/allhome.html

[3] World Resources Institute. Ranking the World's Most Water-Stressed Countries in 2040. Washington, D.C., United States: World Resources Institute; 2015

[4] Belhassan K. Water Scarcity Management in the Maghreb Region. In: Eyvaz et al., editors. Drought - Impacts and Management. London: IntechOpen; 2022. p. 2022. DOI: 10.5772/intechopen.103788

[5] Droogers P, Immerzeel W, Terink W, Hoogeveen J, Bierkens MFP, Van Beek LPH, et al. Water resources trends in Middle East and North Africa towards 2050. Hydrology and Earth System Sciences. 2012;**16**:3101-3114. DOI: 10.5194/hess-16-3101-2012

[6] World Bank. Beyond Scarcity: Water Security in the Middle East and North Africa. MENA Development Report. Washington, DC: World Bank. Licence: CC BY 3.0 IGO; 2017 Available from: https://openknowledge.worldbank.org/handle/10986/27659

[7] Allam MN, Allam GI. Water Resources in Egypt: Future Challenges and Opportunities. Water International. 2007;**32**(2):205-218

[8] MWRI. Facts Regarding the Water Situation in Egypt Report. Ministry of Water Resources and Irrigation Egypt. Cairo: NWRP Project; 2017. Available from: http://gis.nacse.org/rewab/docs/National_Water_Resources_Plan_2017_en.pdf

[9] Kim U, Kaluarachchi JJ. Climate Change Impacts on Water Resources in the Upper Blue Nile River Basin, Ethiopia. Journal of the American Water Resources Association. 2009;**45**(6):1361-1378

[10] Abou Rayan M, Djebedijian B, Khaled I. Evaluation of the effectiveness and performance of desalination equipment in Egypt. Eight. Intern. Water Techn. Conf., IWTC8, Alexandria, Egypt. 2004. pp. 653-668

[11] Abdel-Shafy H, El-Saharty A, Regelsberger M, Platzer C. Rainwater in Egypt: quantity, distribution and harvesting. Mediterranean Marine Science. 2010;**11**(2):245-258. DOI: 10.12681/mms.75

[12] Abo-El-Fadl MM. Possibilities of groundwater pollution in some areas, east of Nile Delta. International Journal of Environment. 2013;**1**(1):1-21

[13] Elnashar WY. Groundwater Management in Egypt. IOSR Journal of Mechanical and Civil Engineering (IOSR-JMCE) e-ISSN: 2278-1684,p-ISSN: 2320-334X. 2014;**11**(4) Ver. IV (Jul- Aug. 2014):69-78

[14] World Bank Group. Climate Change Knowledge Portal. 2021. Available from: https://climateknowledgeportal.worldbank.org/country/algeria

[15] Djellouli Y, Nedjraoui D. Evolution des parcours méditerranéens. In:

Philippe A, Gordon M, editor. Pastoralisme, troupeaux, espaces société. Paris: Hatier. 1995. pp. 449-454

[16] Meddiand M, Boucefiane A. Climate change impact on groundwater in Cheliff-ffZahrez basin (Algeria). Elsevier. APCBEE Procedia. 2013;**2013**(5):446-450

[17] World Bank. Morocco's Economic Outlook- April 2017. 2017. Available from: https://www.worldbank.org/en/country/morocco/publication/economic-outlook-april-2017

[18] MDCE. Morocco's Third National Communication to the Framework Convention on United Nations on climate change (in French). Ministry Delegate to the Ministry of Energy, Mines, Water and the Environment Environmental Officer, April 2016. 2016. Available from: https://unfccc.int/resource/docs/natc/marnc3.pdf

[19] World Bank Group. Managing Urban Water Scarcity in Morocco. Washington, DC: World Bank; 2017 © World Bank. Available from: https://openknowledge.worldbank.org/handle/10986/29190License: CC BY 3.0 IGO

[20] Government of Morocco. National Water Strategy. 2009. Available from: http://www.environnement.gov.ma/PDFs/EAU/STRATEGIE_EAU.pdf

[21] Sinan M, Belhouji A. Impact of the climate change on the climate and the water resources of Morocco on horizons 2020, 2050 and 2080 and measures of adaptation. La Houille Blanche. 2016;**4**:32-39

[22] Ait Kadi M, Ziyad A. Integrated Water Resources Management in Morocco. In: Global Water Security. Vol. 2018. Singapore: Springer; 2018. pp. 143-163

[23] Belhassan K. Relationship Between River and Groundwater: Water Table Piezometry of the Mikkes Basin (Morocco). International Journal of Water Resources and Environmental Sciences. 2020;**9**(1):01-06

[24] Belhassan K. Piezometric Variations of Mikkes Groundwater Basin (Morocco). Research Journal of Earth Sciences. 2011;**3**(2):45-49

[25] Belhassan K. Relationship between River Flow, Rainfall and Groundwater pumpage in Mikkes Basin (Morocco). Iranian Journal of Earth Sciences. 2011;**3**(2):98-107

[26] Belhassan K. Drought - piezometric level of the groundwater aquifers: Mikkes basin (Morocco), Fourteenth International Water Technology Conference, IWTC 14 2010, Cairo, Egypt. 2010. pp. 25-41

[27] Belhassan K, Hessane MA, Essahlaoui A. Interaction between surface water and groundwater in the Mikkes basin Morocco (in French). Hydrological Sciences Journal. 2010;**55**(8):1371-1384. DOI: 10.1080/02626667.2010.528763

[28] World Bank Group. Climate Change Knowledge Portal. 2021. Available from: https://climateknowledgeportal.worldbank.org/country/tunisia/climate-data-historical

[29] Mitchell TD, Hulme M, New M. Climate data for political areas. Area. 2002;**34**:109-112 Available from: http://www.cru.uea.ac.uk/~timm/data/index-table.html

[30] TSNCUNFCCC. 2013. Tunisia's Second National Communication Under the United Nations Framework Convention on Climate Change (UNFCCC). p. 174

[31] Gasmi I, Eslamian S, Moussa M. Evaluation of Traditional Rainwater

Harvesting Technique of "Jessour" in Southern Tunisia, a Case Study on El-Jouabit Catchment. Available from:2021. DOI: 10.1002/9781119776017.ch13

[32] Meddi M, Eslamian S. Uncertainties in Rainfall and Water Resources in Maghreb Countries Under Climate Change. In: Oguge N, Ayal D, Adeleke L, da Silva I. (eds). African Handbook of Climate Change Adaptation. Springer, Cham. 2021. Available from: https://doi.org/10.1007/978-3-030-45106-6_114

[33] Sarra A. Projections, impacts et stratégies d'adaptation. 2009. Available from: https://www.webmanagercenter.com/2009/08/26/79540/tunisie-changement-climatique-projections-impacts-et-strategies-dadaptation-3-3/

[34] Keskes T, Zahar H, Ghezal A. Nexus Assessment for Tunisia. 2019. Available from: https://uploads.water-energy-food.org/legacy/nexus_summary_web_version.pdf

[35] Ageena IM. Trends and patterns in the climate of Libya (1945-2010) [thesis]. Department of Geography and Planning, School of Environmental Sciences. University of Liverpool; 2013. pp. 1-403

[36] Van Camp M, Mtoni YE, Mjemah IC, Bakundukize C, Walraevens K. Investigating seawater intrusion due to groundwater pumping with schematic model simulations: The example of the Dar Es Salaam coastal aquifer in Tanzania. Journal of African Earth Sciences. 2014;**2014**(96):71-78

[37] Seidel SJ, Rachmilevitch S, Schütze N, Lazarovitch N. Modelling the impact of drought and heat stress on common bean with two different photosynthesis model approaches. Environmental Modelling & Software. 2016;**81**:111-121. DOI: 10.1016/j.envsoft.2016.04.001

[38] Sarant L. Egypt. Space to grow. Nature. 2017;**544**:S14-S16. DOI: 10.1038/544S14a

[39] Mohamed A, El-Nahrawy. Country Pasture/Forage Resources Profiles - Egypt. Agricultural Research Centre. Ministry of Agriculture & Land Reclamation, Giza, Egypt; 2011. Available from: https://docplayer.net/47325686-Country-pasture-forage-resource-profiles-egypt-by-dr-mohamed-a-el-nahrawy.html

[40] DGRE. Directorate General of Water Resources (in French). Ministry of Agriculture and Hydrological Resources of Tunisia 43 Street MANOUBIA -TUNIS; 2004. Available from: http://www.hydrosciences.fr/Sierem/Bibliotheque/biblio/annales/TN/2003-2004.pdf

[41] Ghoneim E, Dorofeeva A, Benedetti M, Gamble D, Leonard L, AbuBakr M. Vegetation Drought Analysis in Tunisia: A Geospatial Investigation. Journal of Atmospheric & Earth Sciences. 2017;**1**:002

[42] Salem OM. Management of shared groundwater basins in Libya. African Water Journal. 2007;**2007**(1):109-120

[43] Shahin M. Hydrology and Water Resources of Africa, Ground-water Resources of Africa. Vol. 2003. New York, USA: Kluwer Academic Publishers; 2003, 2003. p. 529

[44] Abu-Zeid K, Elrawady M. 2030 Strategic vision for treated wastewater reuse in Egypt. In: Water Resources Management Program - CEDARE; 2014. p. 48. Available from: http://web.cedare.org/wp-content/

uploads/2005/05/2030-National-Vision-for-Wastewater-Re-use-in-Egypt.pdf

[45] Muamar A, Zouahri A, Tijane M, El Housni A, Mennane Z, Yachou H, et al. Evaluation of heavy metals pollution in groundwater, soil and some vegetables irrigated with wastewater in the Skhirat region "Morocco". Journal of Materials and Environmental Science. 2014;**2014**(5):961-966

[46] Takouleu JM. Reuse of treated wastewater: North Africa and SUEZ set an example. In: Afrik 21Paris, France; 2021. Available from: https://www.afrik21.africa/en/reuse-of-treated-wastewater-north-africa-and-suez-set-an-example/

[47] World Bank. World Reuse in the Arab World from Principle to Practice. A Summary of Proceeding Expert Consultation. Dubai-UAE: Wastewater Management in the Arab World; 2011. p. 2011

[48] Dairi S, Mrad D, Bouamrane A, Djebbar Y, Abida H. A Review on the Reuse of Treated Wastewater in Algeria: Scenario and Sustainability Issues. In: New Prospects in Environmental Geosciences and Hydro-geosciences. Springer International Publishing; 2022. Available from: https://link.springer.com/book/10.1007/978-3-030-72543-3

[49] Salama Y, Chennaoui M, Sylla A, Mountadar M, Rihani M, Assobhei O. Review of wastewater treatment and reuse in the Morocco: Aspects and perspectives. International Journal of Environment and Pollution Research. 2017;**2**(1):9-25

[50] ONAS. Annual report. Tunis: Ministry of Environment and Sustainable Development. (The World Bank Group, 2021). Climate Change Knowledge Portal. 2009. Available from: https://climateknowledgeportal.worldbank.org/country/algeria

[51] Brika B. Water Resources and Desalination in Libya: A Review. Proceedings. 2018;**2018**(2):586. DOI: 10.3390/proceedings2110586

[52] Batisha AF. Water desalination industry in Egypt. In: Eleventh International Water Technology Conference, IWTC11 2007 Sharm El-Sheikh, Egypt. 2007. Available from: https://www.academia.edu/26539414/Water_Desalination_Industry_in_Egypt

[53] Abdel-Shafy HI, Mansour MSM. Overview on water reuse in Egypt: Present and Future. Sustainable Sanitation Practice. 2013;**2013**(14):17-25

[54] Tigrine Z, Aburideh H, Abbas M, Hout S, Kasbadji MN, Zioui D, et al. Membrane Desalination Technology in Algeria: Reverse Osmosis for Coastal Areas. In: Aloui F, Dincer I, editors. Exergy for A Better Environment and Improved Sustainability 2. Cham. Available from:: Green Energy and Technology. Springer; 2018. DOI: 10.1007/978-3-319-62575-1_15

[55] Tahri K. Desalination experience in Morocco. ScienceDirect. Desalination. 2001;**136**(1-3):43-48

[56] Holiday L. Agricultural Water Management: Proceedings of a Workshop in Tunisia. National Research Council, Policy and Global Affairs, Science and Technology for Sustainability Program. Washington, DC: The National Academies Press; 2007. DOI: 10.17226/11880

[57] Mukhtar A, Ghurbal SM. Economics of seawater desalination in Libya. Desalination. 2004;**2004**(165):215-218

[58] Belhassan K. Water scarcity management. In: Vaseashta A, Maftei C,

editors. Water Safety, Security and Sustainability. Advanced Sciences and Technologies for Security Applications. Cham: Springer; 2021. DOI: 10.1007/978-3-030-76008-3_19

[59] Gado TA, El-Agha DE. Feasibility of rainwater harvesting for sustainable water management in urban areas of Egypt. Environmental Science and Pollution Research. 2020;**27**:32304-32317. DOI: 10.1007/s11356-019-06529-5

[60] Dodson L, Bargach J. Harvesting Fresh Water from Fog in Rural Morocco: Research and Impact Dar Si Hmad's Fogwater Project in Aït Baamrane. Procedia Engineering. 2015;**V(107)**:186-193

Chapter 3

The Unique Approaches to Water Management for Transforming Bangladesh's Drought-Prone Northwest Region into a Lush and Granary Landscape

Mohammad Iquebal Hossain and Mohammad Niamul Bari

Abstract

Bangladesh's Barind Tract, lying in the country's northwestern part, is a drought-prone water scarce area that has suffered substantial difficulties in water management for agriculture, drinking, residential, and other uses. The situation has been changed by the efforts of the Barind Multipurpose Development Authority (BMDA). So, the focus of this research is on the numerous initiatives of the BMDA to transform the arid-like Barind area into a green and granary landscape through efficient water management. To achieve this goal, various data sets about water resources development and management, as well as other necessary information were collected from the BMDA and other sources and analyzed. Irrigation was initiated using groundwater (GW) through the installation of deep tube wells (DTWs). DTWs located near the localities are also used to provide drinking water to rural people. Then, by re-excavating derelict ponds, *kharis* (canals), and other water bodies and constructing cross-dams (check dams) and rubber dams in the re-excavated kharies and rivers, surface water (SW) augmentation is started, mostly for supplementary irrigation. Conserved water develops the environment and enhances groundwater recharge (GWR) alongside irrigation. Constructed dug wells in the severely water-stressed areas having no sources of SW and GW supply irrigation for low-water-consuming crops. Pre-paid metering in the irrigation management system has minimized the overuse of water, while the underground pipe water distribution system has reduced water transportation and evaporation losses. The application of managed aquifer recharge (MAR) model helps enhance GWR. Finally, BMDA's efforts have transformed the Barind Tract, as well as Bangladesh's northwest region, into a lush and granary terrain.

Keywords: Barind Tract, groundwater, rainwater, surface water, water management

1. Introduction

The Barind Tract includes parts of the greater Rajshahi, Pabna, and Bogura districts of the Rajshahi Division, as well as parts of the greater Rangpur and Dinajpur districts of the Rangpur division in Bangladesh, as well as parts of the Uttar Dinajpur, Dakhin Dinajpur, and most of Maldah District in West Bengal, India. The Barind Tract in Bangladesh's northwest is the country's largest Pleistocene terrace, made up of Pleistocene alluvial deposits, often known as older alluvium, covered by reddish-brown, sticky Pleistocene silt; Madhupur Clay, with an elevation above sea level ranging from 14 to 45 m [1–6]. The area is geomorphologically separated into three geological units: (a) Barind clay residuum, which overlies and formed on Pleistocene alluvium; (b) Holocene ganges flood-plain alluvium; and (c) Ganges active channel deposits and main distributaries (modern alluvium).The floodplain and the Barind Tract are both physiographically included in the area, which is tectonically part of the Bengal basin's stable shelf region. Level Barind, high Barind, and north eastern Barind are the three primary portions of the Tract [6, 7]. The area's central section is relatively high and irregular. About 80% of the land in high Barind is terraced or undulated. The Tract is located between the latitudes of 24°20'N and 25°35'N, and the longitudes of 88°20'E and 89°30'E. In comparison to other areas of Bangladesh, the Barind Tract's hard red soil and typical dry environment with comparatively high temperatures and low rainfall are crucially significant [8]. The temperature ranges from 8 to 44 degrees Celsius throughout the area [2]. Although the national average for annual rainfall is 2500 mm, the Tract receives lower rainfall ranging from 1250 to 2000 mm, with about 80% of it falling between June and October [9, 10]. The Barind Tract is the most drought-prone and water-scarce region of Bangladesh, with very limited sources of surface water [11, 12]. The area is experiencing groundwater drought, a type of hydrological drought, as well as agricultural drought caused by weather events. In the years 1972, 1975, 1979, 1982, 1986, 1989, 1992, 1994, 2003, 2005, 2009, and 2010, it endured moderate to severe agricultural droughts with hydrological droughts [13]. Because the land is almost flood-free, rainwater is the only means of recharging groundwater [9, 14]. However, rainwater percolation to the aquifer is restricted by more than 15 m thick top clay (Barind clay) layers [15–17] and limited infiltration capacity (2–3 mm/day) [18] reduces natural groundwater recharging [19]. Before 1986, the typical landscape of this region was sun-burned hot-tempered high and low ground with cactus, babla (acacia), herb, and palm trees scattered about [7] and surface water supplies were scarce, and most ponds, canals, rivers, and other natural water bodies dried up during the dry season, leaving crop production entirely dependent on rainfall. Here, only rain-fed T. Amon (Transplanted Amon) (local name of paddy) crops were grown whose cultivation was impeded by a lack of timely rainfall, and field crops were frequently damaged. After T. Amon was harvested, the land remained uncultivated for the remainder of the year due to a lack of water and was used as cattle pasture and people did not have any work with their hands at the time. So, they had to migrate to other parts of the country in search of work. There was an acute shortage of drinking water. For drinking and domestic usage, women used to collect water from open water bodies where cattle were also washed. As a result, many people had to suffer from various water-borne diseases, such as cholera, diarrhea, and dysentery [8, 10, 17]. When the ponds dried up during the dry season, they experienced a major setback in obtaining water to drink. People in the area are familiar with the motto "Barind is a land where life is written in water" [3]. As a result, life was extremely hard, and the area was woefully underdeveloped. In response to the people's and area's vulnerable situation, the Government of Bangladesh approved the "Barind Integrated Area Development Project (BIADP-I)" under the Bangladesh Agriculture Development Corporation

(BADC) in 1986 and, thus the development of the area was started. The project area included three districts: Rajshahi, Nagaon, and C. Nawabganj [20]. The major goal of the project was to boost crop production through the creation of irrigation facilities and to contribute to food security. However, due to administrative difficulties and a delay in the release of funds by BADC to the project authority, the project's progress was hampered. In 1992, Barind Multipurpose Development Authority (BMDA) was established under the Ministry of Agriculture to address this issue and ensure the project's seamless operation. The government approved BIADP-II, and the BMDA successfully implemented it on time. In 2003, the Gov. handed over 1217 unused deep tube wells (DTW) from then north Bengal DTW project of the Bangladesh Water Development Board (BWDB) under Thakurgaon, Dinajpur, and Panchogar districts, recognizing the BMDA's achievement. Within a year, BMDA had successfully put all of the DTWs into operation [21]. Finally, in 2018, the Bangladesh government enacted an act that expanded the BMDA's jurisdiction to include all 16 districts of the Rajshahi and Rangpur Division. **Figure 1** depicts the sequential expansion of BMDA's jurisdiction.

The main goal of BMDA is to turn its land into a granary by creating irrigation facilities. So, the aim of this study is to discuss the water resources development and management activities of BMDA and evaluate them.

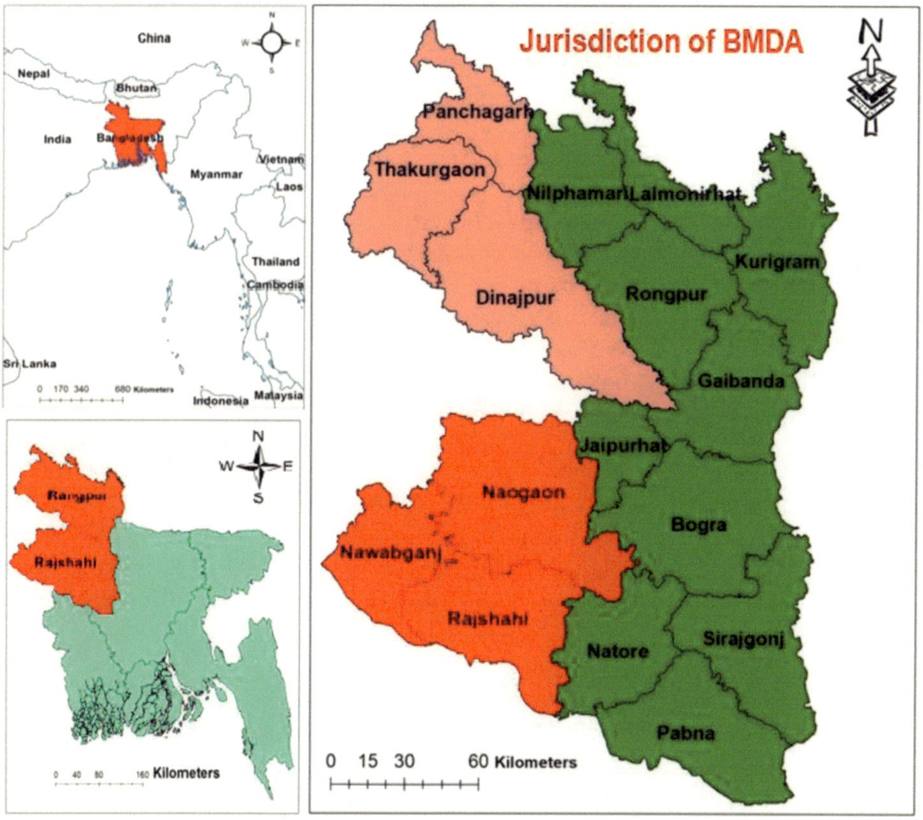

Figure 1.
The map of the sequential expansion of BMDA's jurisdiction into 16 districts of Rajshahi and Rangur division of Bangladesh.

2. Materials and methods

Information on deep tube wells (DTWs) for groundwater extraction and irrigation, re-excavated canals, ponds, cross dams (check dams), rubber dams for surface water augmentation, dug well and recharge well for rainwater harvesting and groundwater recharge, buried pipelines, drinking water supply installations, prepaid meter for irrigation water management, and other pertinent information were collected from the Barind Multipurpose Development Authority (BMDA), and other sources. Different studies were also reviewed. Finally, all of the data and information were checked for accuracy and consistency before being processed in the appropriate format.

3. Results and discussion

3.1 Groundwater irrigation

Because surface water sources are limited in the Barind Tract, irrigation was started with groundwater through the installation of DTWs. There was no way to collect water without DTWs since the aquifer was so deep in the ground (more than 15 m below the surface). Groundwater is now the primary source of irrigation, drinking, and other uses. Deep tube wells were initially run by diesel engines, which caused plenty of problems for farmers, including crop damage. DTW was then electrified for smooth operation, with a submersible motor pump replacing the diesel engine. Some of the irrigation water recharges groundwater through percolation and some go back into the atmosphere, which helps to improve the environment. **Figure 2** shows a DTW in a crop field.

3.2 Derelict kharies and pond re-excavation and rainwater irrigation

Derelict *kharies* (canals) ponds and *beels* (natural water body larger than pond) were re-excavated to augment surface as well as rainwater water by increasing their water holding capacity. For conserving rainwater, cross dams/check dams are constructed at different sections of the re-excavated canal maintaining a certain gradient, depth, and width. Conserved water is mainly used for supplementary irrigation. It also helps to develop an environment and enhance groundwater recharge (GWR). **Figure 3** shows the rainwater conservation system by the re-excavated canal, pond, and *beel*.

Figure 2.
A deep tube well in a crop field with buried pipelines (source: [22]).

Electricity-driven or solar-powered low lift pumps (LLPs) is used to withdraw the conserved water for irrigation (**Figure 4**).

3.3 Irrigation using river water

A floating pontoon with the necessary centrifugal pumps is placed on the river. Pumps are used to draw water from the river and discharge it to the re-excavated canal through an underground pipeline. Water is then withdrawn from the canal by LLP to irrigate the crop fields. As indicated in **Figure 5**, it is a double-lifting irrigation system.

This system of pontoons on the rivers Padma, Mohanonda, and Atrai lessens groundwater exploitation by 7.08% for the respected Upazilas' area [29].

Figure 3.
Surface water augmentation by (a) re-excavated canal, (b) cross dam (check dam) in the re-excavated canal, (c) re-excavated pond and (d) re-excavated natural beel utensils (source: [22]).

Figure 4.
(a) Irrigation by electricity-driven and (b) solar-powered LLP utensils (source: [22]).

Figure 5.
Double lifting irrigation system: (a) Pontoon with centrifugal pumps floating on the river and (b) discharging river water to the canal (source: [29]).

Figure 6.
A rubber dam was constructed across the river (source: [12]).

The irrigation with river water is also operated by constructing a rubber dam across the river to conserve water upstream (**Figure 6**). The LLPs are used to irrigate the crop fields. A rubber dam built across the river Barnai in Puthia Upazila reduces groundwater withdrawals by 10.11% [29].

3.4 Buried pipeline water distribution system

Every irrigation scheme has buried pipelines for conveying water from the pumping station to the agricultural field. It is suitable for the undulated terraced Barind land. Water can be conveyed from a lower elevation to a higher elevation with this system and conveyance and evaporation losses in the pipeline part are minimized. Because the pipeline runs beneath the earth's surface, valuable land is saved. The system is shown in **Figure 7**.

3.5 Dug Well (DW) irrigation for severely water-stressed area

A dug well is typically built in a severely water-stressed environment with no sources of surface water or aquifer that produces groundwater. By a dug well, low water consuming crops, such as tomato, cauliflower, chilies, eggplant, and other

Figure 7.
Buried pipeline system: (a) pipe laying stage, (b) header tank for raising pressure head, and (c) outlet with air vent (source: [22]).

Figure 8.
A dug well in a vegetable field (source: [12]).

vegetables, can be cultivated. During rainfall, rainwater (RW) is collected by the solar panel cum rainwater harvesting device and stored in the dug well. A solar-powered pump lifts the stored water to an overhead tank; it is then fed to the crop area for irrigation. The details of DW are shown in **Figure 8**.

3.6 Prepaid metering in irrigation management

To avoid the overuse of valuable water, BMDA has introduced a prepaid metering system in irrigation management. Each farmer has a prepaid card (user card) that can be recharged by paying a certain amount of money to the BMDA local office or recruited dealer. When a farmer needs irrigation water for his crop field, he inserts his card into the irrigation equipment's prepaid meter, such as DTW or LLP. If the card has a balance, the irrigation pump automatically starts and distributes irrigation water. The

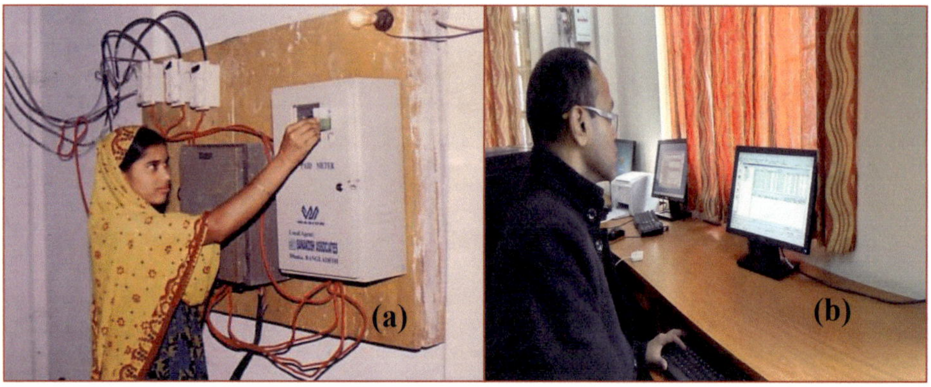

Figure 9.
Prepaid metering system: (a) a woman inserts her prepaid card into the irrigation equipment's prepaid meter, and (b) the system is monitored over the internet network from the BMDA's headquarters (source: [22]).

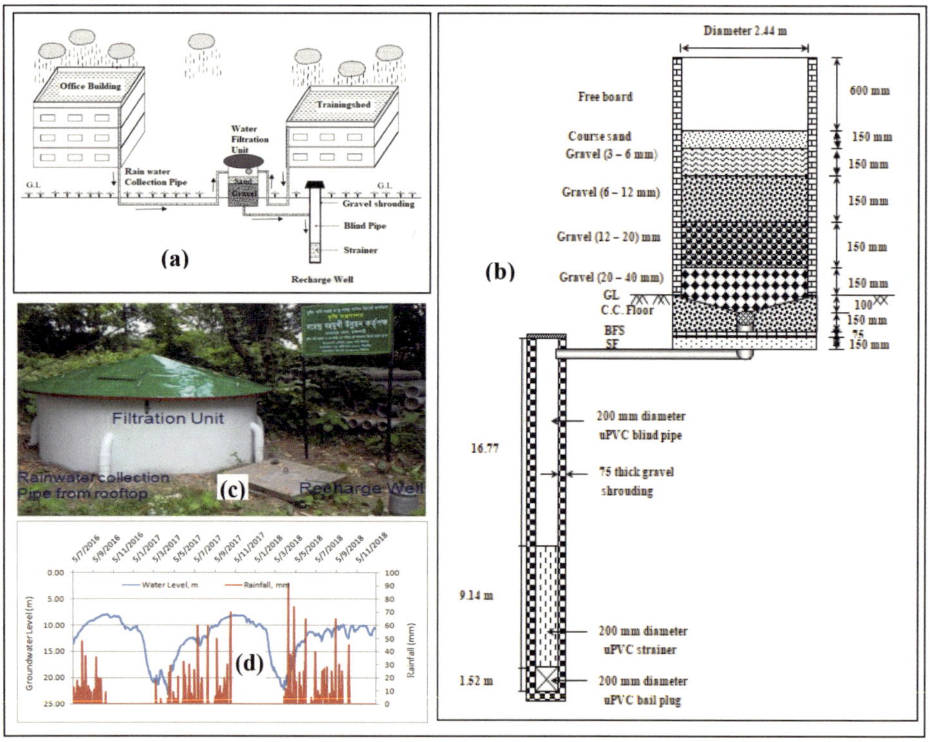

Figure 10.
BMDA Mohanpur office campus MAR model: (a) schematic view of whole MAR model, (b) longitudinal sectional image of FU and RW, (c) photo of FU with RW, and (d) GWL hydrograph illustrating recharge performance with rain water infiltration (source: [12]).

DTW automatically turns off when the balance on the inserted card is zero or the card is removed from the meter. Farmers only use irrigation water for the actual needs of crop fields because it is a prepaid system. As a result, by lowering irrigation expenses, the over-use of irrigation water is reduced. The irrigation actions of this system can be monitored centrally through the internet using a digital method. Now, for all farmers of an irrigation scheme, it is possible to cultivate crops with less water. **Figure 9** depicts the system.

3.7 Application of MAR model to enhance groundwater recharge

In 2016, at the BMDA Mohanpur zonal office site, a recharge well (RW) with a water filter unit (FU) was built as a modified MAR model taking into account the lithology, as well as the aquifer condition and maximum daily rainfall (**Figure 10**). The uPVC pipeline collects rooftop rainwater from the office building and a training shed, filters it via the FU's sand-gravel filtration media, and then recharges the groundwater through RW. To monitor water level fluctuations, an observation well (OW) with an auto water level recorder (AWLR) system has been installed. The rainfall data is recorded at a rain gauge station. The water level may now be automatically monitored from the BMDA headquarters in Rajshahi via the internet network every day of the year, which is the pioneer attempt in Bangladesh [17].

3.8 Drinking water supply

There was an acute shortage of drinking water (DW) in this area. So, finding no way, people had to collect pond water for drinking, domestic and other usages faced multifarious water-borne diseases like cholera, diarrhea, dysentery, etc. About 15% of irrigation wells of BMDA are located near the villages. A 25,000 liter capacity overhead water tank with a 2400 m pipeline network is constructed to supply drinking water to the rural people from this irrigation well alongside irrigation. **Figure 11** shows the drinking water supply system in detail.

Through the analysis of collected data and different studies following observations can be mentioned:

- The buried uPVC pipeline system has a mean conveyance efficiency of between 94.46 and 95.37%, with a rate of water loss of between 5.45 and 6.11%, and about 80% of farmers commend it [23]. For the 15525 DTW schemes, a total of 2000 hectares of land have been saved [24].

Figure 11.
(a) An overhead water tank close to an irrigation DTW, (b) women are collecting water for drinking purposes, and (c) using the system's water, a woman is washing utensils (source: [22]).

- About 0.52 million hectares of land have been brought under-regulated irrigation facilities with the installation of 15525 DTWs and 13,512 km buried pipelines, benefiting 0.96 million farmers [25].
- With 3357 re-excavated ponds, 2063 km of re-excavated canal, 747 cross dams built at various portions of the re-excavated canal, 11 floating pontoons with required pumps set up on the river Padma, Mohanonda, and Atrai, and a rubber dam, 97,000 hectares of land have been possible to irrigate. To irrigate crop fields, 601 electricity-driven LLP and 168 solar-powered LLP are used. A total of 171,000 farmers are being benefited from the current system [25].
- The installation of 572 dug wells has enabled the irrigation of low-water-use crops on 1620 hectares of land, benefiting around 11,000 farmers [25].
- Due to the construction of 1579 drinking water supply infrastructure, about 1,250,000 rural people now have access to potable water [25].
- Every year, around 300,000 liters of rainwater replenish groundwater from the MAR infrastructure established at the BMDA Mohanpur office [17].
- The installation of a prepaid meter system in irrigation management reduced the overuse of valuable irrigation water and resulted in a 22% increase in the command area [26]. The private irrigation system (PIS) has an average irrigation cost, that is, 112% higher than the prepaid meter irrigation system (PMIS), and the PIS uses 39% of the extra water [27]. This research shows that the prepaid irrigation scheme is a cost-effective and ecologically friendly way to increase rice production in the study area [27].
- Since 1986, cropping intensity has increased from 113 to 230% [3, 28, 29].
- As a result of BMDA's efforts to develop and manage water resources, single-cropped dry land has been turned into triple-cropped land, making it green and granary.

4. Conclusions

Due to the augmentation of groundwater by 15525 DTWs along with 13,512 km buried pipeline, 0.52 million hectares of land are now under groundwater irrigation benefiting 0.96 million farmers. Due to the augmentation of surface water by 3357 re-excavated ponds, 2063 km of the re-excavated canal, 747 cross dams, 11 pontoons, and a rubber dam, 97,000 hectares of land have been brought under surface water irrigation, benefiting 171,000 farmers. When compared to surface channel, buried pipeline increased conveyance efficiency from 94.46 to 95.37%, saving 2000 hectares of land at the 15525 DTW schemes. The introduction of prepaid meters has reduced irrigation water usage, resulting in a 22% increase in the command area. About 1,250,000 rural people receive potable water from 1579 drinking water installations.

Conserved water from re-excavated water bodies and irrigation water from crop fields assist replenish groundwater and enhance the environment in tandem with irrigation, which requires extensive research to determine the exact amount of recharge. A single MAR can replenish groundwater with over 300,000 l of rainwater. As a consequence of BMDA's innovative endeavors in water resource augmentation and management for diverse reasons, the land has been turned into a lush and granary environment. To improve groundwater recharge, other office buildings, training sheds, educational institution buildings, commercial buildings, and residential houses should be placed under the managed aquifer recharge program. More canals ponds and other derelict water bodies should be re-excavated to conserve additional surface water and utilize more river water, and they must be maintained on a regular basis to ensure their long-term viability. For sustainable usage, future planning and management of water resources in the area, a comprehensive assessment of groundwater, and surface water resources, including zoning, is required.

Author details

Mohammad Iquebal Hossain[1,2]* and Mohammad Niamul Bari[1]

1 Deptartment of Civil Engineering, Rajshahi University of Engineering and Technology, Bangladesh

2 Barind Multipurpose Development Authority, Rajshahi, Bangladesh

*Address all correspondence to: iquebal_hossain@yahoo.com

References

[1] Ahmed KM. Barind Tract. Bangladesh: National Encyclopedia of Bangladesh; 2006. Available from: http://en.banglapedia.org/index.php?title=Barind_Tract

[2] Ahmeduzzaman M, Shantanu K, Asad A. A study on ground water fluctuation at Barind Area, Rajshahi. International Journal of Engineering Research and Applications (IJERA). 2012;**2**(6):1465-1470

[3] Barind Multipurpose Development Authority (BMDA). Barind Authority: Past-Present. 2006

[4] Matin I. Groundwater of barind tract-its condition and potentiality. In: 33rd Annual Convention of Institution of Engineers. Bangladesh; 1989

[5] Khandoker RA. Origin of elevated Barind Madhupur areas, Bengal Basin: Results of neotectonic activities. Bangladesh Journal of Geology. 1987;**6**:1-9

[6] IWM. Groundwater Model Study for Deep Tubewell Installation Project in Barind Area, Final Report. Vol. III2006

[7] Barind Multipurpose Development Authority (BMDA). Final Report on Barind Environmental Action Plan (BEAP). 2003

[8] Barind Multipurpose Development Authority (BMDA). Development Activities of BMDA. Bangladesh: Rajshahi; 2021

[9] IWM. Groundwater Resources Study and Decision Support System Development of Rajshahi, Naogaon, Chapai Nawabganj, Pabna and Natore Districts and Also Remaining District of Rajshahi Division through Mathematical Model Study for Barind Integrated Area Development Project, Phase-III, Final Report. Vol. I2012

[10] Rashid MA, Hossain MI. Irrigation management system of BMDA: An innovative approach. In: Proceedings of 2nd International Conference on Planning, Architechture & Civil Engineering. Rajshahi; 2019

[11] Bari N, Hossain MI, Miah SU. Development of storm water pretreatment system for managed aquifer recharge in water-stressed Barind Tract. Arabian Journal of Geosciences. 2021;**14**:2071. DOI: 10.1007/s12517-021-08398-0

[12] Hossain MI, Bari N, Miah SU. Opportunities and challenges for implementing managed aquifer recharge models in drought-prone Barind tract, Bangladesh. Applied Water Science. 2021;**11**:181. DOI: 10.1007/s13201-021-01530-1

[13] Rahman ATMS, Jahan CS, Mazumder QH, Kamruzzaman M, Hosono T. Drought analysis and its implication in sustainable water resource management in Barind Area, Bangladesh. Journal Geological Society of India. 2017;**89**:47-56

[14] Rahman M, Mahbub AQM. Groundwater depletion with expansion of irrigation in Barind Tract: A case study of Tanore Upazila. Journal of Water Resource and Protection. 2012;**4**:575-567. DOI: 10.4236/jwarp.2012.48066

[15] Hossain MI, Bari N, Miah SU. Operational constraints in conventional managed aquifer recharge in Barind Area at North-Western Region of Bangladesh. In: Procedings of 2nd International Conference on Planning, Architechture & Civil Engineering, Faculty of Civil

Engineering, Rajshahi University of Engineering & Technology. Rajshahi; 2019

[16] Hossain MI, Bari MN, Miah SU, Jahan CS, Rahaman MF. Performance of MAR model for stormwater management in Barind Tract, Bangladesh. Groundwater for Sustainable Development. 2020; **10**(2020):100285. DOI: 10.1016/j.gsd.2019.100285

[17] Hossain MI, Bari MN, Miah SU, Kafy AA, Nasher NMR. Application of modified managed aquifer recharge model for groundwater management in drought-prone water-stressed Barind Tract, Bangladesh. Environmental Challenges. 2021;**4**(2021):100173. DOI: 10.1016/j.envc.2021.100173

[18] Jahan CS, Mazumder QH, Adham MI, Hossain MMA, Haque A-M. Study on groundwater recharge potentiality of Barind Tract, Rajshahi District, Bangladesh using GIS and remote sensing technique. Journal of Geological Society of India. 2010;**75**:432-438

[19] Jahan CS, Rahman ATMS, Mazumder QH, Kamruzzaman M. Adaptation for climate change effect on groundwater resource through MAR technique in Drought Prone Barind Area, Rural Bangladesh. In: Ali SM, editor. Bangladesh: Combating Land Degradation and Drought. Dhaka: Ministry of Environment (MoEF); 2015

[20] Imon AHMR, Ahmed M. Water level trend in Barind area. Malaysian Journal of Mathematical Sciences. 2013;**7**:1-15

[21] Barind Multipurpose Development Authority (BMDA). Development Activities of BMDA. Bangladesh: Rajshahi; 2021

[22] Hossain MI, Bari MN, Miah SU, Reza MS, Jahan CS, Hasan MR, et al. IWRM approaches in water resources management activities of BMDA, Barind Tract of Bangladesh: A blooming effort. Hydrology. 2020;**8**:91-97

[23] Rahman M, Kamal AHM, Miah SU. Study on the irrigation water distribution system developed by Barind Multipurpose Development Authority. Journal of the Bangladesh Association of Young Researchers (JBAYR). 2011;**1**:63-71

[24] Barind Multipurpose Development Authority (BMDA). Development Project Proposal of "Command Area Extension and Development Project". 2010

[25] Barind Multipurpose Development Authority (BMDA). Annual Report: 2020-2021. 2021

[26] Asaduzzaman. Replication of the Barind model to Muhuri irrigation project. Moving from water problems to water solutions: Research needs assessment for the Eastern Gangetic Plains. In: Proceedings of the International Workshop held at the National Agricultural Science Complex (NASC), Indian Council of Agricultural Research (ICAR). New Delhi, India; 2013

[27] Kashem MA, Rahman MS. Feasibility of Prepaid irrigation at Rajshahi region in Bangladesh. Technical Journal of River Research Institute. 2018;**14**(1):29-34

[28] Barind Multipurpose Development Authority (BMDA). Annual Report: 2019-2020. 2020

[29] Hossain MI, Bari MN, Kafy AA, Rahman ZA, Rahman MT. Application of double lifting method for river water irrigation in the water stressed Barind Tract of northwest Bangladesh. Groundwater for Sustainable Development. 2022. DOI: 10.1016/j.gsd.2022.100787

Chapter 4

Remote Sensing for Agricultural Applications

Saju Adhikary, Benukar Biswas, Manish Kumar Naskar, Bishal Mukherjee, Aditya Pratap Singh and Kousik Atta

Abstract

The application of remote sensing in quantifying the crop health status is trending. Sensors can serve as early warning systems for countering climatic or biological aberrations before having negative impacts on crop yield. Remote sensing applications have been playing a significant role in agriculture sector for evaluating plant health, yield and crop loss (%) estimation, irrigation management, identification of crop stress, weed and pest detection, weather forecasting, gathering crop phenological informations etc. Forecast of crop yields by using remote sensing inputs in conjunction with crop simulation models is getting popular day by day for its potential benefits. Remote sensing reduces the amount of field data collection and improves the precision of the estimates. Crop stress caused by biotic and abiotic factors can be monitored and quantified with remote sensing. Monitoring of vegetation cover for acreage estimation, mapping and monitoring drought condition and maintenance of vegetation health, assessment of crop condition under stress prone environment, checking of nutrient and moisture status of field, measurement of crop evapotranspiration, weed management through precision agriculture, gathering and transferring predictions of atmospheric dynamics through different observational satellites are the major agricultural applications of remote sensing technologies. Normalized difference vegetation index (NDVI), vegetation condition index (VCI), leaf area index (LAI), and General Yield Unified Reference Index (GYURI) are some of the indices which have been used for mapping and monitoring drought and assessing vegetation health and productivity. Remote sensing with other advanced technologies like geographical information systems (GIS) are playing a massive role in assessment and management of several agricultural activities. State or district level information systems based on available remote sensing information are required to be utilized efficiently for improving the economy coming from agriculture.

Keywords: remote sensing, agriculture, vegetation indices, yield forecast

1. Introduction

In remote sensing, objects or areas of the real world are studied by gathering information at a distance without physical contact. In addition to ground

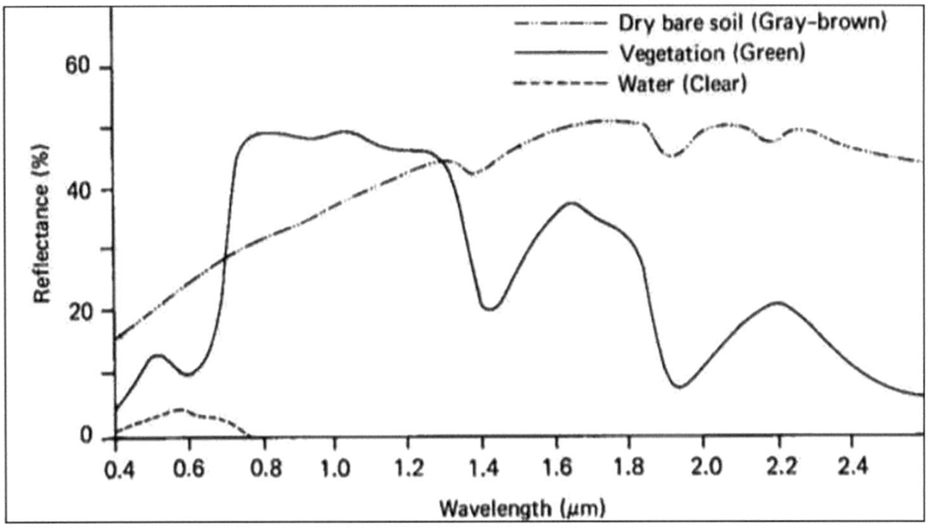

Figure 1.
Typical spectral reflectance curves for vegetation, dry bare soil and water.

observations, remote sensing uses non-contact technologies to monitor the earth's resources. The spectral characteristics of different objects are unique in nature and can be utilized to derive information such as temperature, water content etc. The use of visible, infrared and microwaves to evaluate any physical features has been well established. It is used to distinguish vegetation from bare soil, water, and other similar features based on the responses of the targets to these wavelength regions (see **Figure 1**). Also, it can be used in monitoring crop growth, land use pattern and land cover changes, mapping of water resources and monitoring of water status, weather forecasting and crop yield estimation, and monitoring diseases and pest infestations. Thus the application of remote sensing data in agriculture can provide a timely, efficient, and cost efficient approach [1]. Several agrometeorological applications are also possible. It is very useful to forecast crop yields by using remote sensing inputs in conjunction with crop simulation models. For complementing traditional meteorological and crop status data collection methods, the space based satellite technology is becoming increasingly important, since ground and air-based platforms are time consuming and limited in their use.

2. Agricultural applications

Satellite remote sensing began with most researchers using data for land cover classification, with farmers focusing on crop types as a major application. Plant biophysical properties have become more important in agricultural remote sensing in recent years. The use of remote sensing in agriculture has been around for a long time. The classification of crop canopies based on image processing is one the biggest milestone achieved in this field. In precision agriculture, remote sensing offers the advantage of providing repeated information without destructive sampling of crops, which can provide useful information. In large geographic areas, remote sensing provides an inexpensive alternative to traditional data collection methods [2]. Agriculture crop acreage and production are mainly estimated by satellite remote sensing in India. Based on biophysical attributes of crops and/or soils, remote sensing technology has the potential to revolutionize agricultural productivity detection and characterization [3]. Using satellite data, yield estimation can be done [4, 5], crop phenological information can be gathered [6], stress situations can be identified [7] and disturbances can also be detected. As a result of the combined use of remote sensing and GIS, spatial variables of interest can be created that can be applied to a variety of fields, including flood plain mapping, hydrological modeling, surface energy flux, urban development, land use changes, crop growth monitoring, and stress detection [8]. Increasing spatial resolution of aircraft or satellite mounted sensors has led to advances in remote sensing methods that use narrow band or hyperspectral sensors. A more detailed analysis of crop classification has also been enhanced by hyperspectral remote sensing. Using a combination of principal component analysis, lambda-lambda models, stepwise discriminant analyses, and a derivative greenness vegetation index, Thenkabail et al. [9] conducted rigorous analysis on hyperspectral sensors (between 400 and 2500 nm) in order to determine whether there had been any change in crop composition. There have been many investigation procedures using sensors which can provide reliable data in a timely manner at a fraction of the cost of traditional data gathering methods.

3. Monitoring of vegetation cover

Agricultural acreage estimation and yield assessment rely heavily on the science of remote sensing. Researchers used aerial photographs as well as digital image processing techniques to conduct a number of experiments. However, remote sensing reduces the amount of field data collected and improves the precision of estimates [8]. A significant improvement in crop characterization, discrimination, modeling, and mapping is known to be possible with hyperspectral data, compared with broadband multispectral remote sensing [10]. Using 33 optimal HNBs and an equal number of two-band normalized difference HVIs, Thenkabail et al. [10] characterized, categorized, mapped, and studied biophysical and biochemical quantities of major agricultural crops. Remote sensing techniques are generally used to assess the crop's health and yield based on physical parameters such as nutrient stress and water availability. The spectral characteristics of any vegetation depends on various factors such as water content, cell density, elemental concentration etc. This feature is used to extract numerous information of a sample field from different type of spectral band. Remote sensing indices are being used by other researchers to provide synoptic perspectives on regional crop conditions. Rouse et al. [11] proposed the Normalized Difference Vegetation Index as a way to assess vegetation condition [11]. There has been a great deal of effort made to develop additional vegetation indices that reduce the influence of soil background and atmosphere on spectral measurements, as the NDVI has become the most widely used vegetation index [12, 13]. Remotely sensed vegetation data can be managed with SAVI (Soil Adjusted Vegetation Index), an index developed by Huete [14]. A number of indices have been used for mapping and monitoring drought and assessing vegetation health and productivity, including the normalized difference vegetation index (NDVI), vegetation condition index (VCI), leaf area index (LAI), and General Yield Unified Reference Index (GYURI) [4, 15, 16]. A very high resolution radiometer (AVHRR) data was used by Kogan et al. [17] to model corn yield and early drought warning in China using vegetation indices from Advanced Very High Resolution Radiometer (AVHRR) data. A semi-arid region yield and

Index	Formula and spectral bands or wavelengths (nm)	Level/sensor	Application
Advanced normalized vegetation	$ANVI = \frac{NIR-BLUE}{NIR+BLUE}$	Airborne(RMKTOP 15camera)	Mapping Rodolfo segetum
Index	**BLUE:400–500** **NIR: 700–900**		Patches in sunflower crop
Aphid Index		Ground based (ASD FieldSpec3spectrometer)	Identification of aphid infestation in mustard
	RED1:712 RED2:719		
	NIR1:761 NIR2:908		
Chlorophyll	$CI = \frac{NIR}{GREEN} - 1$	Ground based	Plant nitrogen
Index	**GREEN:520–600** **NIR: 760–900**	(Exotic radiometer)Satellite (Quick Bird)	Status estimates
Effective	$ELAI = -0.441 + 0.285\frac{NIR}{RED}$	Ground based	Winter oilseed
Leaf Area Index		(CIMEL313 radiometer)	rape yield prediction
	RED:610–680		
	NIR: 780–890		
Green	$GNDV1 = \frac{NIR-GREEN}{NIR+GREEN}$		
Normalized	**GREEN:557–582**	Airborne	Corn yield
Difference	**NIR: 720–920 and/or**	(Multispectral	predictions
Vegetation	**GREEN:520–600**	Digital	
Index	**NIR: 760–900**	Camera)	
Green Red	$GRV1 = \frac{GREEN-RED}{GREEN+RED}$	Ground based	Estimation of
Vegetation Index	**GREEN:520–590**	(GER1500 Spectro radiometer)	Damage caused by thrips
	RED:620–680		
Healthy	$HI = \frac{GREEN-RED1}{GREEN+RED1} 0.5RED2$	Airborne	Early
Index	**GREEN:534**	(MCA-6 and Micro-Hyper spec	detection of Verticillium wilt of olive

Table 1.
Some examples of vegetation indices having specific applications in agricultural sector.

irrigated wheat distribution is estimated using leaf area indices from four satellite scenarios [18]. As shown in **Table 1**, there are some vegetation indices that can be used specifically for agricultural purposes.

4. Crop condition assessment

The use of remote sensing to assess plant health through spectral information can be beneficial in agriculture. It is possible to detect stress in plants by using remote sensing techniques due to physiological changes caused by stress [19]. Monitor crop growth at regular intervals in order to take appropriate actions and also to determine whether any stress factor is likely to affect production. A variety of factors contribute

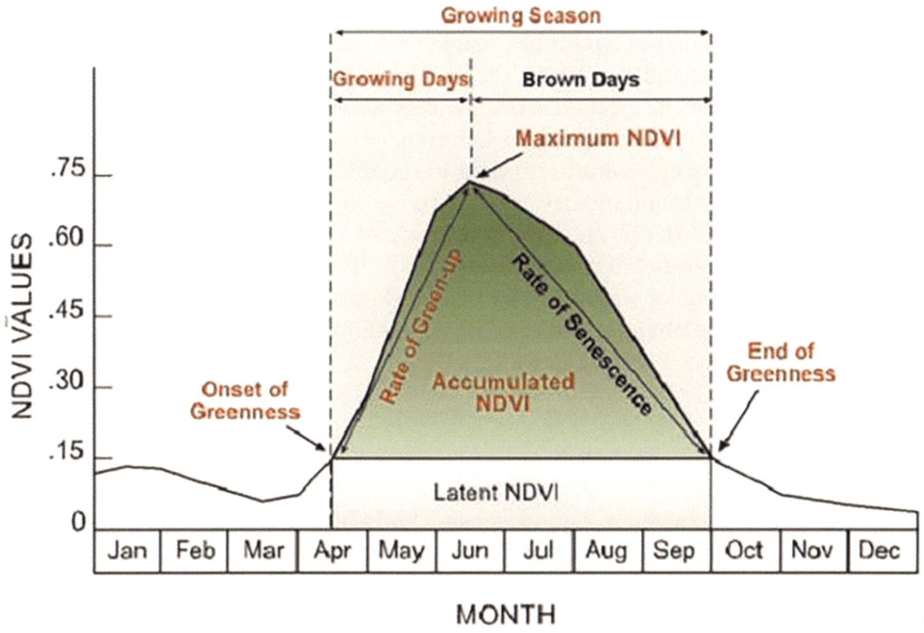

Figure 2.
A diagram of a hypothetical 12-month NDVI multi-temporal vegetation response curve for native vegetation that is typical of the Great Plains region, USA. Shown on the graph are selected vegetation phenology metrics that can be extracted through the analysis of the NDVI, near cloud-free datasets (adopted from Reed et al. [22]).

Type	Metric	Interpretation
Temporal	1. Time of onset of greenness	Beginning of photosynthetic activity
	2. Time of end of greenness	End of photosynthetic activity
	3. Duration of greenness	Length of photosynthetic activity
	4. Time of maximum greenness	Time when photosynthesis at maximum
NDVI-value	5. Value of onset of greenness	Level of photosynthesis at start
	6. Value of end of greenness	Level of photosynthesis at end
	7. Value of maximum NDVI	Level of photosynthesis at maximum
	8. Range of NDVI	Range of measurable photosynthesis
Derived	9. Accumulated NDVI	Net Primary Production (NPP)
	10. Rate of green–up	Acceleration of increasing photosynthetic activity
	11. Rate of senescence	Acceleration of decreasing photosynthetic activity
	12. Mean daily NDVI	Mean daily photosynthetic activity

Table 2.
Vegetation phenology metrics characterize vegetation phenology and are used to develop summary regional data for research on agro-ecosystem attributes (after Reed et al. [22]).

to the growth stages and development of crops, including soil moisture, date of planting, air temperature, and day length. The conditions and productivity of plants are influenced by these factors. Too high temperatures at pollination, for example, can negatively impact corn crop yields. Forecasters may be able to better predict corn yields if they know the temperature when pollination occurs [20]. Siddiqui [21] explains that drought makes land inhospitable to humans, livestock, biomass potentials, and plant species, and also causes the land to be incapable of cultivation. Drought monitoring through satellite data has been widely accepted now and the Vegetation Condition Index (VCI) and Normalized Difference Vegetation Index (NDVI) are widely used to identify agricultural drought in regions with different ecological conditions. Many vegetation indices are used to measure crop growth and condition, such as reflectance ratios, NDVIs, PVIs, transformed vegetation indexes, and greenness indices. With operational remote sensing, NDVI profiles are extracted each year for 12 Vegetation Phenology Metrics (VPMs), and these metrics are used to analyze agricultural vegetation changes due to changes in climate and land management practices (**Figure 2** and **Table 2**) [22].

5. Nutrient and water status

Through precision farming, remote sensing and GIS can be applied to nutrient and water stress management, which are the most important fields. Utilizing remote sensing and GIS to detect nutrient stress can help us reduce cultivation costs and increase fertilizer efficiency for crops through site-specific nutrient management. Precision farming technologies can be used to judiciously use water in semi-arid and arid regions. Das and Singh [23] demonstrated that drip irrigation combined with remote sensing data can improve the efficiency of water use by reducing runoff and percolation losses. In the visible region, water stressed crops displayed higher spectral reflectance than non-stressed crops. There was a difference between stressed and non-stressed crops in terms of vegetation indices like NDVI, RVI, PVI, and GI. In the field, soil moisture can be estimated using microwave remote sensing. Through remote sensing data, it is possible to obtain information on crop water demand, water use, soil moisture conditions, and related crop growth at various stages. Sri Lankan irrigation projects were assessed using NOAA satellite data by Bandara [24], for example. In this analysis, irrigation efficiency was determined by comparing estimates from remote sensing with actual water availability. Based on high resolution land data assimilation system (HRLDAS), Das et al. [25] developed a soil moisture and temperature map for India with a spatial resolution of 1 km, in near real-time (with a few hours' latency) for four soil depths and vegetation root zones. Remote sensing has played an important role in understanding crop soil characteristics with the development of hyperspectral bands in the thermal region. Precision farming can be more effective with such information provided in conjunction with GPS. It has been demonstrated that nitrogen leaching occurs more often in wet tropical and subtropical climates due to spatial variability of soil properties, such as SOM content [26], water content [27] and yield zones [28, 29]; these properties are having a direct impact on the N nutrition status of corn plants. Bredemeier and Schmidhalter [30] indicate that this results in the overfertilization of some sites and the under fertilization of others. By using crop sensors [31, 32], we are able to increase nitrogen fertilization efficiency with variable-rate nitrogen fertilization (VRF).

6. Crop evapo-transpiration

Increasing temperatures and irregular rainfall cause a decrease in soil moisture, which in turn decreases crop productivity. It is defined as a situation in which precipitation and evapo-transpiration are balanced over the long-term average, which is also affected by the timing and potency of the monsoon [33]. The relationship between water stress and a plant's thermal characteristics is described by vegetation indices such as CWSI (Crop Water Stress Index) [34], ST (Surface Temperature) [35], WDI (Water Deficit Index) [36], and SI (Stress Index) [37]. Based on MODIS data, Sruthi et al. [38] calculated NDVI values and correlated them with land surface temperatures for the Raichur district of Karnataka. In conjunction with the vegetation index, the LST provides early warning systems to farmers if a region is experiencing an agricultural drought. Evapo-transpiration estimates are essential for evaluating irrigation scheduling, calculating water and energy balances, determining crop water stress indexes (CWSIs), and determining climatological and meteorological conditions. As soil water availability and crop evapo-transpiration are directly related to plant temperature, the energy emitted by cropped areas has been useful in assessing crop water stress. The AVHRR and MODIS data can be used in estimating evaporative fraction (EF) demonstrated by Batra et al. [39]. The spatio-temporal extent of agricultural drought in Rajasthan state was monitored using NOAA-AVHRR NDVI data by Dutta et al. [40]. Using airborne remote sensing to measure crop evapotranspiration, Neale et al. [41] provide an overview of crop coefficients obtained from high resolution airborne remote sensing. There are various approaches to calculating evapo-transpiration from remote sensed data; most use simple correlations between remote sensed data and evapo-transpiration, but some combine different types of remotely sensed data. Water management for agricultural systems relies heavily on remote sensing. By developing hyperspectral sensors and integrating the remote sensing data with other spatial data, this can be further enhanced.

7. Weed identification and management

In order to manage weeds effectively, precision weed management techniques are helpful. Precision agriculture combined with remote sensing is today's most promising technology. Although ground surveying for mapping weeds is very labour intensive and time-consuming, it is a good method for mapping weed information for a specific location. Image-based remote sensing can be utilized for weed detection and site-specific weed management [36, 42, 43]. Using remote sensing technology to identify weeds in crops and to develop weed maps in the field can allow site specific and need-based herbicides to be applied to manage weeds based on the difference in spectral reflectance properties between weeds and crops. The radiance ratio and the NDVI values of solid stands of wheat and solid weed plots were higher in Kaur et al. [44]. Beyond 30 days, radiance ratios and NDVI can be used to distinguish pure wheat from pure *Rumex spinosus* populations. Rumex populations at different levels could be discriminated between themselves after 60 DAS. The radiance ratio and NDVI were used by Kaur and Jaidka [45] to distinguish pure wheat from pure populations of *Malva neglecta* after 30 days after planting and to remain distinct up to 120 days later, and to discriminate between different levels of weed populations after 60 days of planting. Farmers can be advised to take preventive measures based on weed prescription maps that can be prepared with Geographic Information System (GIS).

8. Pest and disease infestation

Crop stress caused by biotic and abiotic factors can be monitored and quantified with remote sensing. To prevent the spread of insects and take effective control measures, remote sensing methodologies need to be perfected for identifying insect breeding grounds. Using remote sensing to assess and monitor insect defoliation, spectral response to chlorosis, yellowing leaves, and foliage reductions over a given period of time has been used to relate those differences' correlations, classifications, and interpretations [46]. Lee et al. [47], for example, have applied remote sensing to map and detect defoliation, characterize pest destruction pattern, and provide information to pest management decision support systems. The authors of William et al. [48] analyzed Landsat imagery before and after defoliation to determine which vegetation types were healthy and unhealthy. A study conducted by Debeurs and Townsend [2] concluded that MODIS data could be used for estimating vegetation indices in plots and determining insect-damaged defoliation. Using remote sensing technology to identify pest-infested and diseased plants has proven to be an effective and inexpensive method reported by Riedell et al. [49]. For detecting specific insect pests and identifying disease damage on oat, they used remote sensing techniques. It is suggested that remote sensing can be used to measure canopy characteristics and spectral reflectance differences in oat crop canopies in order to assess insect infestation damage and disease infection damage. Wheat Streak Mosaic disease management in the wheat crop can be supported by accurate detection and quantification of disease using the Landasat 5TM image, according to Mirik et al. [50]. Using multispectral remote sensing, Franke and Menz [51] concluded that fungal wheat diseases can be monitored with high resolution.

9. Crop yield and production forecasting

Various statistical-empirical relationships between yield and vegetation indices have been used in remote sensing to forecast crop yields [52, 53]. Information on crop production before harvest is essential for national food policy planning. Forecasting crop production requires reliable crop yields. A number of factors influence crop yield, including crop variety, water and nutrient status of a field, weed infestations, pest and disease infestations, and weather parameters. It is dependent on these factors that the spectral response curve appears. Spectral response curves indicate the crop's performance and condition based on their growth and decay. Menon [19] suggests that growth profiles can be constructed and yield related parameters can be extracted by using IRS P3 WiFS (Wide Field Sensor) and IRS-1C WiFS.

10. Precision agriculture

As scientists, engineers and large-scale crop growers increasingly use remote sensing technology for precision farming [3], it is becoming a key component of the same. With the help of the sensors fitted in farm machines, precision farming aims to reduce the cost of cultivation, improve control, and improve resource utilization efficiency. One of the most advanced components of precision farming is variable rate technology (VRT). The moving farm machines contain sensors with a computer that recommends inputs based on GPS data. This allows the application of inputs to be controlled based on input recommendation maps [54]. Using precision farming, you can make management decisions based on information acquired at frequent intervals and at high spatial resolutions. In order to provide such information, remote sensing is undoubtedly an important tool. Multispectral remote sensing was used by Bagheri et al. [55] to manage nitrogen fertilizer at specific sites. An Iranian corn-planting area measuring 23 ha was imaged with the advanced spaceborne thermal emission and reflection radiometer (ASTER).

11. Atmospheric dynamics

The use of meteorological satellites in weather forecasting is one of the other applications of remote sensing. Cloud cover, wind, moisture, temperature, and wind speed are measured by meteorological satellites. It may be possible to determine whether there is enough or inadequate water in the field based on variations in canopy temperature. It is possible to use canopy temperature variability (CTV) to monitor crop water stress [19] as well as canopy air temperature difference (CATD) to determine if the canopy is overwatered [19]. The use of remote sensing data for drought assessment plays a significant role in agriculture. NDVI produced by NOAA-AVHRR data is used to assess and monitor droughts at the district level, allowing timely preventive and corrective measures to be taken.

12. Future prospects

Even at small farms, remote sensing is highly useful in detecting and managing various crop problems, including abiotic and biotic stresses. State or district level information systems based on available remote sensing and GIS crop information are required to efficiently utilize the information on crops for improving the economy. In order to adopt policies or address national issues related to agriculture, governments can make use of remote sensing data. Plant and seed tissue nano-chips can be implanted in near-real time to monitor crops using a new and non-traditional remote sensing technique. In the future, remote sensing will clearly play a more important role in assessing agricultural science.

13. Conclusion

The resolution of the satellite images (MODIS- minimum resolution 250 m, LANDSAT- minimum resolution 15 m) are the key constraints in this field. Agricultural fields are so much versatile in nature with respect to moisture content, topography, elevation, nutrient content that the empirical relationship established between source to sink is highly vulnerable. Standardized relationship between the crop spectral characteristics and physio-chemical properties needs to be established crop specifically. The initial installation cost of a satellite is till now very high and out of scope for an individual. The lone application of remote sensing in precision agriculture may lead to confusion and wastage of resources so the importance of ground truthing still remain pertinent. This makes the application of satellite derived remotely sensed data quite complicated and costly for the policy makers. The recent advancement in high resolution field level image processing obtained through drones compiled with the satellite derived knowledge is quite promising. Though the remote sensing applications have a lot of theoretical consideration and backdrops, it has a huge scope in agricultural sciences to serve the humanity with higher production, vulnerability prediction and impact assessment under the context of rapid climate change.

Acknowledgements

The author would like to thank to the co-authors for their valuable inputs.

Conflict of interest

The authors declare no conflict of interest.

Author details

Saju Adhikary[1], Benukar Biswas[1], Manish Kumar Naskar[2], Bishal Mukherjee[1], Aditya Pratap Singh[3] and Kousik Atta[4]*

1 Department of Agronomy, Bidhan Chandra Krishi Viswavidyalaya, Nadia, West Bengal, India

2 Department of Agricultural Meteorology and Physics, Bidhan Chandra Krishi Viswavidyalaya, Nadia, West Bengal, India

3 Department of Genetics and Plant Breeding, Bidhan Chandra Krishi Viswavidyalaya, Nadia, West Bengal, India

4 Department of Plant Physiology, Bidhan Chandra Krishi Viswavidyalaya, Nadia, West Bengal, India

*Address all correspondence to: kousikatta1995@gmail.com

References

[1] Justice CO, Townshend JRG, Vermata EF, Masuoka E, Wolfe RE, Saleons N, et al. An overview of MODISL and at a processing and product status. Remote Sensing of Environment. 2002; **83**:3-15

[2] de Beurs K, Townsend P. Estimating the effect of gypsy moth defoliation using MODIS. Remote Sensing of Environment. 2008;**112**:3983-3990

[3] Liaghat S, Balasundram SK. A review: The role of remote sensing in precision agriculture. American Journal of Agricultural and Biological Sciences. 2010;**5**(1):50-55

[4] Doraiswamy PC, Sinclair TR, Hollinger S, Akhmedov B, Stern A, Prueger J. Application of MODIS derived parameters for regional crop yield assessment. Remote Sensing of Environment. 2005;**97**:191-202

[5] Bernardes T, Meriera MA, Adami M, Giarolle A, Rudorff BFT. Monitoring biennial bearing effect on coffee yield using MODIS remote sensing imagery. Remote Sensing. 2012;**4**:2492-2509

[6] Sakamoto T, Yokozawa M, Toritani H, Shibayama M, Ishitsuka N, Ohno H. A crop phenology detection method using time series MODIS data. Remote Sensing of Environment. 2005; **96**:366-374

[7] Gu Y, Brown JF, Verdin JP, Wardlow B. A five-year analysis of MODISNDVI and NDWI for grassland drought assessment over the central great plains of the United States. Geophysical Research Letters. 2007;**34**: 1-6

[8] Kingra PK, Majumder D, Singh SP. Application of remote sensing and GIS in agriculture and natural resource management under changing climatic conditions. Agricultural Research Journal. 2016;**53**(3):295-302

[9] Thenkabail PS, Enclona EA, Ashton MS, Van Der Meer B. Accuracy assessments of hyperspectral wave band performance for vegetation analysis applications. Remote Sensing of Environment. 2004;**91**(3–4):354-376

[10] Thenkabail PS, Mariotto I, Gumma MK, Middleton EM, Landis DR, Huemmrich KF. Selection of hyperspectral narrow bands (HNBs) and composition of hyperspectral two band vegetation indices (HVIs) for biophysical characterization and discrimination of crop types using field reflectance and hyperion/EO-1Data. IEEE Journal of Selected Topics in Applied Earth Observations and RemoteSensing. 2013;**6**(2):427-439

[11] Rouse JW, Haas RH, Schell JA, Deering DW. Monitoring vegetation systems in the Great Plains with ERTS. In: Third ERTS Symposium, NASA SP-351. 1973. pp. 309-317

[12] Calvao T, Palmeirim JM. Mapping Mediterranean scrub with satellite imagery: Biomass estimation and spectral behaviour. International Journal of Remote Sensing. 2004;**25**: 3113-3126

[13] Wallace JF, Caccetta PA, Kiiveri HT. Recent developments in analysis of spatial and temporal data for landscape qualities and monitoring. Australian Ecology. 2004;**29**:100-107

[14] Huete AR. A soil-adjusted vegetation index (SAVI). Remote Sensing of Environment. 1988;**25**:295-309

[15] Ferencz CS, Bognar P, Lichtenberger J, Hamar D, Tarcsai GY, Timar G, et al. Crop yield estimation by satellite remote sensing. International Journa lof Remote Sensing. 2004;**25**(20): 4113-4149

[16] Prasad AK, Chai L, Singh RP, Kafatos M. Crop yield estimation model for Iowa using remote sensing and surface parameters. International Journal of Applied Earth Observation and Geo information. 2006;**8**(1):26-33

[17] Kogan F, Yang B, Guo W, Pei Z, Jiao X. Modelling corn production in China using AVHRR-based vegetation health indices. International Journal of Remote Sensing. 2005;**26**(11):2325-2336

[18] Hadria R, Duchenin B, Lahrouni A, Khabba S, Er-Raki S, Dedieu G, et al. Monitoring of irrigated wheat in a semi-arid climate using crop modelling and remote sensing data: Impact of satellite revisit time frequency. International Journal of Remote Sensing. 2006;**27**(6): 1093-1117

[19] Menon ARR. Remote sensing applications in agriculture and forestry. In: A paper from the Proceedings of the Kerala Environment Congress. 2012. pp. 222-235

[20] Nellis MD, Pricey KP, Rundquist D. Remote sensing of crop land agriculture. In: The SAGEH and Book of Remote Sensing. 2009. p. 26

[21] Siddiqui AR. Regional Evaluation of Desertification Hazards in the Arid lands of Western Rajasthan [an unpublished PhD thesis]. Aligarh, Uttar Pradesh, India: AMU; 2004. p. 221

[22] Reed BC, Brown JHF, Vander Zee D, Love Land TR, Merchant JW, Ohnlen DO. Measuring phenological variability from satellite imagery. Journal of Vegetation Science. 1994;**5**: 703-714

[23] Das DK, Singh G. Estimation of evapotranspiration and scheduling irrigation using remote sensing techniques. In: Proc. Summer Inst. On Agricultural Remote Sensing in Monitoring Crop Growth and Productivity. New Delhi: IARI; 1989. pp. 113-117

[24] Bandara KMPS. Monitoring irrigation performance in Sri Lanka with high frequency satellite measurements during the dry season. Agricultural Water Management. 2003;**58**(2):159-170

[25] Das KC, Singh J, Hazra J, Sheraton LC. Development of a high resolution soil moisture for precision agriculture in India. In: Proceedings of the 14th International Conference on Precision Agriculture, Monteral, QC, Canada. 2018. pp. 26-27

[26] Casa R, Cavalieri A, Locascio B. Nitrogen fertilization management in precision agriculture: A preliminary application example on maize. Italian Journal of Agronomy. 2011;**6**:23-27

[27] Delin S, Berglund K. Management zones classified with respect to drought and waterlogging. Precision Agriculture. 2005;**6**:321-340

[28] Blackmore S, Godwin RJ, Fountas S. The analysis of spatial and temporal trends in yield map data over six years. Bioscience & Engineering. 2003;**84**: 455-466

[29] Bramley RGV. Lessons from nearly 20 years of precision agriculture research, development, and adoption as a guide to its appropriate application. Crop and Pasture Science. 2009;**60**: 197-217

[30] Bredemeier C, Schmidhalter U. Laser-induced chlorophyll fluorescence sensing to determine biomass and nitrogen uptake of winter wheat under controlled environment and field conditions. In: Stafford JV, editor. Precision Agriculture, Proceedings. Wageningen: Academic Publishers; 2005. pp. 273-280

[31] Singh I, Srivastava IA, Chandna P, Gupta R. Crop sensors for efficient nitrogen management in sugarcane: Potential and constraints. Sugar Technology. 2006;**8**:299-302

[32] Li Y, Chen D, Walker CN, Angus JF. Estimating the nitrogen status of crops using a digital camera. Field Crops Research. 2010;**118**:221-227

[33] Wilhite DA, Glantz MH. Understanding the drought phenomenon: The role of definitions. Water International. 1985;**10**:111-120

[34] Jackson RD, Idso SB, Reginato RJ, Pinter PJ. Canopy temperature as a crop water stress indicator. Water Resources Research. 1981;**17**:1133-1138

[35] Jackson RD. Remote sensing of biotic and abiotic plant stress. Annual Review of Phytopathology. 1986;**24**:265-286

[36] Moran MS, Inoue Y, Barnes EM. Opportunities and limitations for image-based remote sensing in precision crop management. Remote Sensing of Environment. 1997;**61**(3):319-246

[37] Vidal A, Pinglo F, Durand H, Devaux-Ros C, Maillet A. Evaluation of a temporal fire risk in dexin Mediterranean forests from NOAA thermal IR. Remote Sensing of Environment. 1994;**49**:296-303

[38] Sruthi S, Aslam M, M. A. Agricultural drought analysis using the NDVI and land surface temperature data; a case study of Raichur District. International Conference On Water Resources, Coastal and Ocean Engineering. 2015;**4**:1252-1264

[39] Batra N, Islam S, Venturini V, Bisht G, Jiang L. Estimation and comparison of evapotranspiration from MODIS and AVHRR sensors for clear sky days over the southern Great Plains. Remote Sensing of Environment. 2006; **103**:1-15

[40] Dutta D, Kundu A, Patel NR, Saha SK, Siddiqui AR. Assessment of agricultural drought in Rajasthan (India) using remote sensing derived Vegetation Condition Index (VCI) and Standardized Precipitation Index (SPI). The Egyptian Journal of Remote Sensing and Space Science. 2015;**18**(1):53-63

[41] Neale CMU, Jayanthi H, Wright JL. Irrigation water management using high resolution air borne remote sensing. Irrigation and Drainage Systems. 2005; **19**(3–4):321-336

[42] Johnson GA, Cardina J, Mortensen DA. Site–specific weed management: Current and future direction. In: *The State of Site–Specific Management for Agriculture*. Wiley Pubblisher; 1997. pp. 131-147

[43] Lamb DW, Weedon MM, Rew LJ. Evaluating the accuracy of mapping weeds in seeding crops using air-borne digital imaging: *Avena* spp.in seeding triticale. Weed Research. 1999;**39**(6): 481-492

[44] Kaur R, Jaidka M, Kingra PK. Study of optimum time span for distinguishing *Rumex spinosus* in wheat crop through spectral reflectance characteristics. Proceedings of the National Academy of Sciences, India, Section B: Biological Sciences. 2014;**84**(3):625-633

[45] Kaur R, Jaidka M. Spectral reflectance characteristics to distinguish *Malvaneglectain wheat (Tritium aestivum)*. Indian Journal of AgriculturalSciences. 2014;**84**(10): 1243-1249

[46] Franklin S. Remote Sensing for Sustainable Forest Management. Boca Raton, FL: Lewis Publisher; 2001. p. 407

[47] Lee W, Alchanatis V, Yang C, Hirafuji M, Moshou D, Li C. Sensing technologies for precision specialty crop production. Computer and Electronic inAgriculture. 2010;**74**:2-33

[48] Williams D, Stauffer M, Leung K. A forester's look at the application of image manipulation techniques to lands at data. In: Symposium on Remote Sensing for Vegetation Damage Assessment, February 14–16. Falls Church, VA: Washington The Society; 1979. pp. 221-229

[49] Riedell WE, Osborne SL, Hesler LS. Insect pest and disease detection using remote sensing techniques. In: Proceedings of 7th International Conference on Precision Agriculture. Minneapolis, MNUSA; 2004

[50] Mirik M et al. Remote monitoring of wheat streak mosaic progression using sub-pixel classification of Landsat 5 TM imagery for site specific disease Management in Winter Wheat. Advances in Remote Sensing. 2013;**2**: 16-28

[51] Franke J, Menz G. Multi temporal wheat disease detection by multi spectral remote sensing. Precision Agriculture. 2007;**8**:161-172

[52] Thenkabail PS, Smith RB, De-Pauw E. Evaluation of narrow band and broad band vegetation indices for determining optimal hyperspectral wave bands for agricultural crop characterization. Photogrammetric Engineering. 2002;**68**: 607-621

[53] Casa R, Jones HG. LAI retrieval from multi-angular image classification and inversion of a ray tracing model. Remote Sensing of Environment. 2005;**98**: 414-428

[54] NRC. Precision Agriculture in the 21st Century Geospatial and Information Techniques in Crop Management. Washington, DC: National Academy Press; 1997. p. 149

[55] Bagheri N, Ahmadi H, Alavipanah SK, Omid M. Multispectral remote sensing for site-specific nitrogen fertilizer management. International Agrophysics. 2013;**26**:103-108

Chapter 5

Prediction of the Transported Soil Volume by the Presence of Water in the Vicinity of Ma'adim Vallis (Mars)

Emilio Ramírez-Juidías

Abstract

Ma'adim Vallis is a channel that ends at the Gusev Crater. In general terms, the length of the channel is about 700 km while its width can reach 20 km and its depth 2 km. Currently, the images obtained from the area allow to visualize a landscape of abundant gullies with important signs of water erosion. In order to predict the volume of transported soil by the presence of water in the vicinity of Ma'adim Vallis, as well as to generate a rainfall model applicable to the red planet, a total of 16 cross-sectional profiles were made along the main canyon, ensuring that all were equidistant from each other depending on the orographic characteristics of the study area. Once the volume of transported soil was obtained, a novel model capable of predicting the rainfall concentration index (RCI) necessary to produce a certain water erosion on the Mars surface was obtained. This model is applicable to other rocky planets as a result of its simplicity.

Keywords: DEM, advanced remote sensing, rainfall model in Mars, big data analysis

1. Introduction

Currently, it is believed that the gravitational accretion process, of material from the protoplanetary disk that orbited the Sun, was the reason by creating Mars about 4.6 billion years ago [1, 2]. At that time, the solar wind accumulated the elements with a lower boiling point on the outermost rocky planet, which explains why the red planet has a higher concentration of Cl, P and S than Earth [3]. Regarding its initial orbit, there are studies [4] that support the theory that its formation occurred in the asteroid belt, taking place around 120 million years later, a migration towards its current position. The presence of liquid water on the surface would gradually disappear as a result of the existence of a tenuous atmosphere and an intense solar wind [5].

In another vein, about 4 billion years ago, during the late heavy bombardment (LHB), most of the impact basins were created on planets' surface [6], which it given rise to the dichotomy between both Northern and Southern hemispheres of the red planet [7].

With regard to Martian geology, it is very important to bear in mind that the record of its evolution is preserved in the rocks and sediments existing on the surface of the red planet. In this sense, and according to [8], minerals can fingerprint many processes that build the Martian rock record. For this reason, and with the purpose of knowing the composition of this planet, infrared spectroscopy instruments have been used in order to make a series of mineralogical maps from orbiters or rovers currently existing on the surface.

As is well known, the surface of Mars is made up of ferric oxides and oxyhydroxides, giving its surface its characteristic red color. However, under the dust layer there is a basaltic crust, which bears some similarity to the terrestrial, more specifically to the oceanic. In relation to the above, it must be emphasized that the early Martian crust, formed during the first billion years, preserves a rich chemical record of multiple and diverse hydrated environments. Nevertheless, the more recent crust exhibits a less frequent and less intense interaction between itself and water [8]. This fact can reinforce the theory of the continuous drying of the planet and, therefore, the presence of high concentrations of salts on the surface.

Regarding its atmosphere, Mars is composed mainly of carbon dioxide (95.7% CO_2, 2.1% Ar, 2% N_2, traces of O_2, CO, water and CH_4 among other gases [9]). The mean atmospheric pressure on its surface is 6 mbar, about 0.6% of the mean sea level pressure on Earth. It ranges from a low of 0.30 mbar at the peak of Mount Olympus to more than 11.55 mbar in the depths in Hellas Planitia. Another interesting fact is that the mass of the Martian atmosphere (about 25×10^{15} kg) is about 200 times less than the Earth's (about 5148×10^{15} kg).

According to [10], Mars has important similarities to Earth, such as the presence of polar caps, the existence of seasonal changes, and the presence of observable weather patterns. While the climate of Mars has similarities to that of Earth, including periodic glaciations, there are also important differences, such as much lower thermal inertia. Although there has been a notable increase in sublimation in the polar regions, in recent decades there has been a decrease in global temperature, probably due to the same cyclical phenomenon that exists on Earth and, thanks to which, the climate change makes possible, every thousands of years, the appearance of a thermal variation with great repercussions on the entire planet. The fact of being further from the Sun than the Earth, as well as the existence of a tenuous atmosphere that retains little heat, causes the average surface temperature to be about −55°C, with superficial thermal variations from 27°C in summer, in the equator, down to −143°C in the polar caps [11]. It is convenient to emphasize dust storms (wind speed of around 200 km/h) caused by the difference in energy that the planet receives at aphelion and perihelion. Since they occur on a global scale, they cause a decrease in maximum temperatures (due to the decrease in energy from the Sun), and an increase in minimum temperatures.

In general, and with regard to the orography of the planet, [5] specifies that the existing channels, as well as the gullies and the runoff processes associated with them, suggest the presence of liquid water on the surface in the past. Furthermore, taking into account that the atmospheric pressure in the Martian past was ostensibly higher than the current one, and according to [12], raindrops with enough size could be formed to give rise to orographic incidents on the surface.

Regarding the erosion processes, reference [5] obtained a series of predictive algorithms of soil transported volume valid on Mars surface. Other studies [13] specifies that levitation forces enhance the downslope transport on Mars, even though there is evidence that in the past, in the period in which the atmospheric density was such that it allowed the existence of water in a liquid state on its surface, the loss of soil due to sedimentary transport in watercourses was the force that it modeled the morphology of the Martian surface. In relation to the above, it is necessary to bear in mind the work of reference [14], since one way of being able to study the climatic conditions of Mars (when liquid water existed on the surface) is by observing the value of the angle formed by the valleys network ramifications on Mars. This consideration is due to the fact that, over the years, the angle that these ramifications have usually remains unchanged.

Although numerous studies have been carried out on Mars to date, none of them have obtained the transported soil volume using cross sections methodology. In the same way, no scientific work has inferred an algorithm capable of obtaining a novel model that can predict the rainfall concentration index (RCI) necessary to produce a certain water erosion on the Mars surface. For this reason, this work pretends to be a reference in future studies because this chapter's novel model may be applicable for rocky planets.

2. Study area

Located in the southern hemisphere (**Figure 1**), Ma'adim Vallis (21,98° S; 177,5° E in planetocentric coordinates) is a channel that ends at the Gusev Crater (14,44° S;

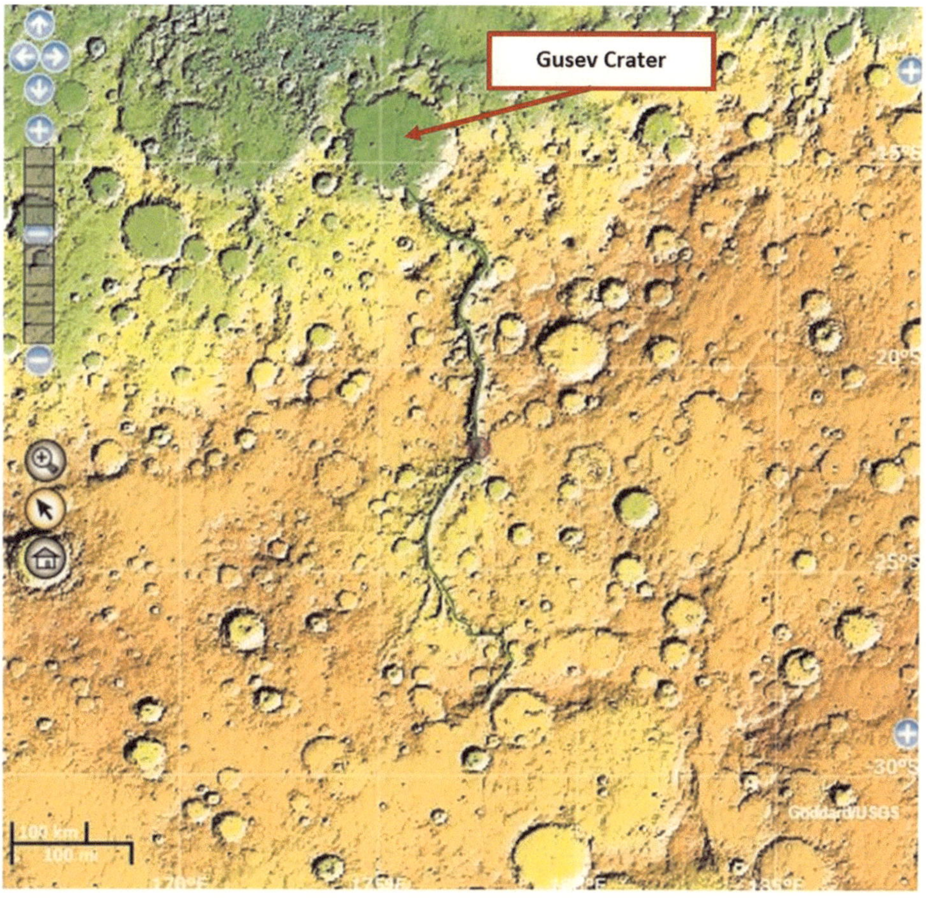

Figure 1.
Location of the study area.

175,29° E also in planetocentric coordinates). In general terms, the length of the channel is about 700 km while its width can reach 20 km and its depth 2 km.

Several studies [15, 16] suggest that Ma'adin Vallis originated as a result of the overflow of a lake located in the Eridania basin. As a result of this event, Ma'adim Vallis has a peculiar topography (**Figure 2**) characterized by a network of channels of variable width (8–25 km), as well as a system consisting of step terraces and several inner channels that flow into the main one.

According to [15], Ma'adim Vallis enters Gusev crater through an opening in its southern edge, and a separate crevasse opens northwestward toward the northern lowlands, suggesting past flow into Gusev crater. It should be noted that the divides of Gusev crater and the Eridania basin are only partially incised, so volumetrically reduced central basins still exist. However, the northern divide of the intermediate basin was downcut to below its floor level, opening the basin completely for through-flowing drainage [16].

On the other hand, and considering the longitudinal profile of Ma'adim Vallis (**Figure 3**), it can be divided into three different areas. The first one, descends from the beginning (445 m high) to the first 324.48 km (−796 m high) with an average slope of 0.38%. This is followed by a 0.17% slope area in 193,820 km, where the elevation difference is 332 m. The last part, with a slope of around 0.21%, has a length of 198.18 km, being the mouth towards the Gusev crater.

In relation to the above, it is necessary to specify that both at the beginning and at the end of the longitudinal profile considered in the present work, the width of the main channel is practically constant, being much greater in the central part, which corresponds to the area of less slope. This fact may be due to the fact that, logically, in the steepest areas, water erosion wears down the channel in depth, while in those with a lower slope value, the erosion occurs mainly in the width of the channel.

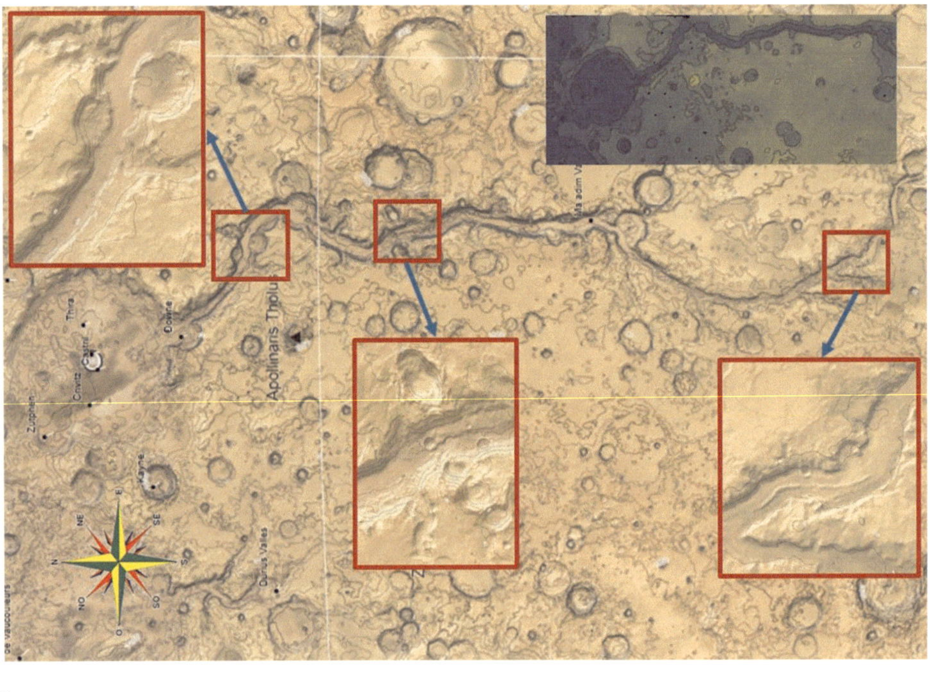

Figure 2.
Topography of Ma'adim Vallis.

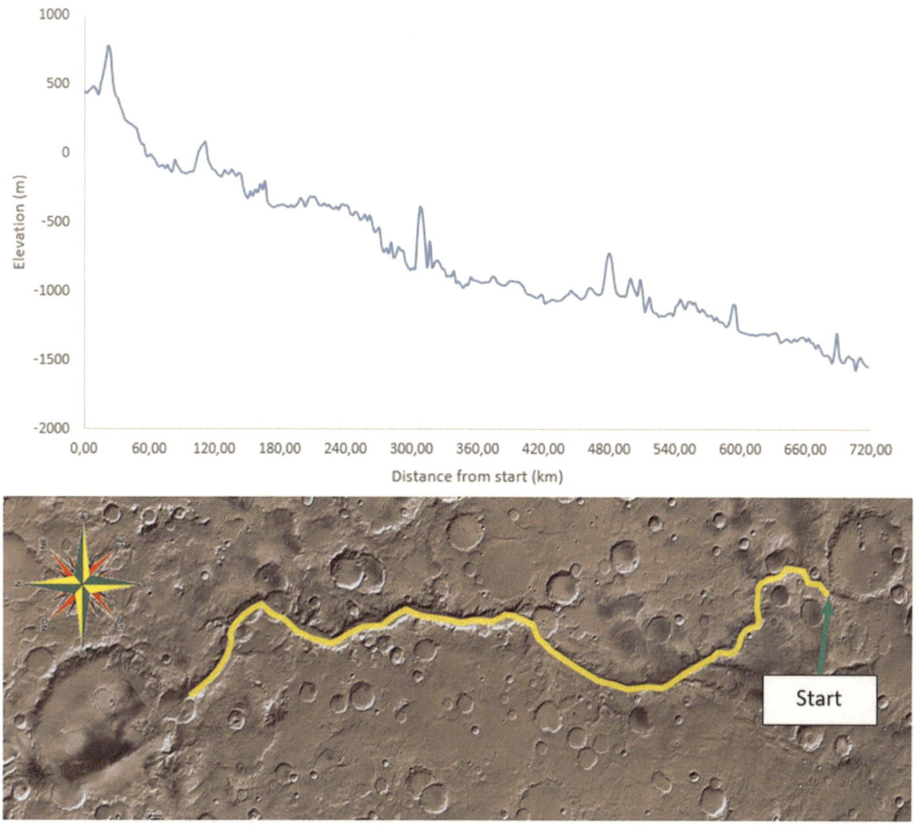

Figure 3.
Longitudinal profile of study area.

As is evident, the causes of water erosion are varied (rainfall, continuous flow of water "river erosion", existence of tides and associated waves, or even occasional dripping on the ground), although the water erosion produced by subsurface runoff should not be forgotten.

Based on the characteristics of the study area, as well as those corresponding to areas of Earth with evidence of water erosion, it is clear that Ma'adim Vallis has been shaped by water.

Perhaps, this is one of the reasons why Ma'adim Vallis is one of the most important areas of Mars, which shows that in the past there was a predominant constant hydrological cycle on the planet, and therefore, more favorable conditions for the existence of life.

3. Materials and methods

For the development of this work, a total of 16 cross-sectional profiles were selected, equidistant among themselves, along the main channel of the study area (**Figures 4** and 5).

In order to obtain the amount of soil lost (eroded), over time, until giving rise to the gully system that predominates in Ma'adim Vallis, it was proceeded to measure, in

Cross-section profile	Top edge width (km)	Base width (km)	Depth (km)
1	11,410	2,992	0,677
2	9,342	2,215	0,710
3	12,042	3,078	0,545
4	15,820	1,356	0,550
5	15,077	5,786	0,590
6	12,139	2,388	0,607
7	10,540	3,260	0,651
8	*27,185*	*11,273*	*1,393*
9	*35,925*	*20,118*	*0,649*
10	18,865	9,894	1,657
11	18,056	6,216	1,731
12	18,615	4,875	1,820
13	18,409	7,699	1,648
14	17,850	9,135	1,390
15	25,193	8,541	1,632
16	*30,510*	*1,695*	*1,828*
Mean ==>	18,561	6,283	1,130
	Gully length ==>	771	km

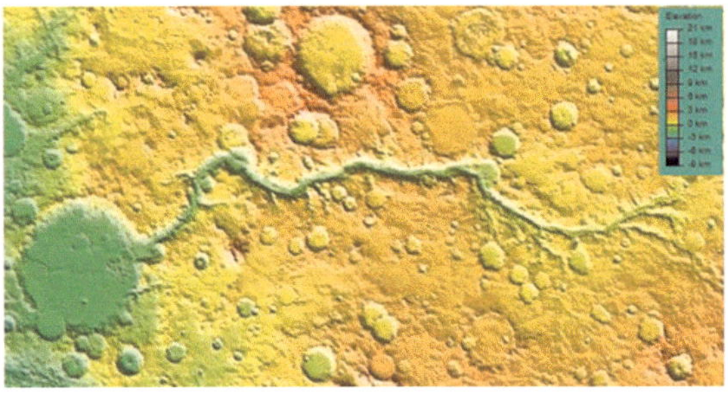

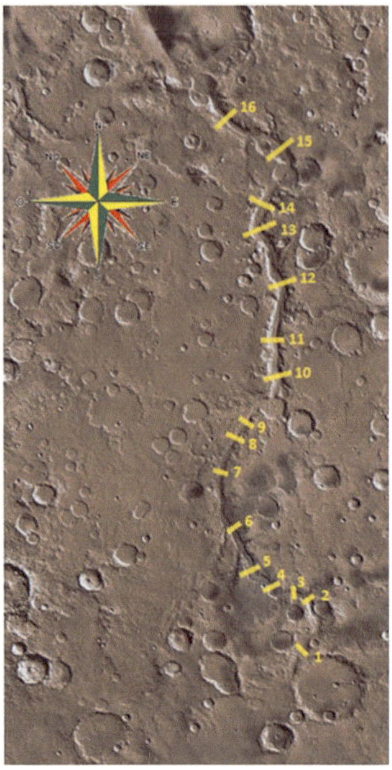

Figure 4.
Location of cross-section profiles, and data, in the study area.

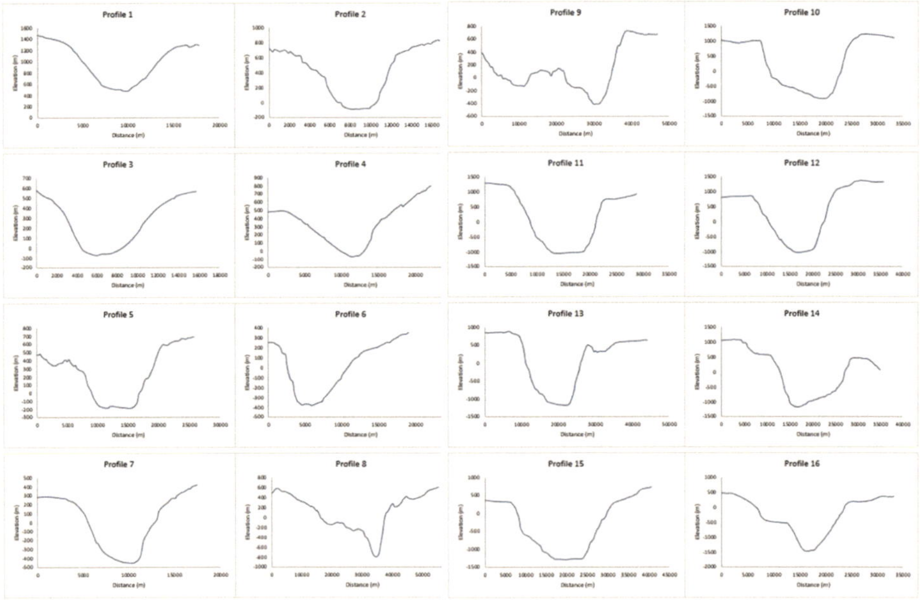

Figure 5.
Cross-section profiles.

each transverse profile, both the upper width "U_w" (in km) and the base width "B_w" (in km) of the gully, as well as its depth "D" (in km). Subsequently, the mean values of the three mentioned variables were calculated, thanks to which, and by applying

Eq. (1) in order to always be on the side of safety, it was possible to obtain the soil lost volume (V) in the study area. Logically, the channel, or gully, length "L" (in km) will be necessary.

$$V\ (\mathrm{km}^3) = [(U_\mathrm{w} + B_\mathrm{w}) \cdot D] \cdot L\ (\mathrm{km}) \tag{1}$$

In a later phase, the data from MOLA was downloaded (the soil lost volume was also determined by using of Topocal 2022-v9.0.811 software) and, after an exhaustive analysis, the contributing watershed area corresponding to the study area was determined. This was necessary to convert the soil lost volume to an equivalent per km^2.

In another vein, a model capable of predicting the RCI through [17] was developed (Eq. 2) taking into account the similar behavior that liquid water has on the surface both on Earth and on Mars.

$$\mathrm{RCI} = e^{\frac{1.56 + \log\left(\frac{Q'_s}{t'}\right) - 0.46 \cdot \log(H) \cdot \tan(S)}{2.65}} \tag{2}$$

where:

Q'_s = relationship between the existing sediments mass (in g) and the contributing watershed area (in m^2) in Ma'adim Vallis.

t' = time required, in Martian years, to generate the sediment mass existing in the study area. It is necessary to specify that $t' = 1.882 \times t$, where "t" is the Earth elapsed time in years.

H = it is the average elevation (in m), taking as 0 the lowest elevation of the longitudinal profile corresponding to the gully base.

S = it is the average slope of the contributing watershed (in °).

Finally, in order to adequately reference this work, an exhaustive bibliographic review was carried out.

4. Results and discussion

The soil loss, and its consequent degradation, can occur in many ways. In general, after a degradation process, the soil becomes thin and stony, giving rise to severe impacts for the surrounding environment, such as those specified below:

- Increase in bulk density, as well as crusting of the upper soil layer.
- Reduction of the upper layer thickness of the soil, in addition to the exhumation of the subsoil as a result of long periods of continuous erosion.
- As a result of the fact that Ma'adim Vallis is located in the tropical zone (orange area "zone D") of Mars (**Figure 6**), it must be taken into account that an increase in the acidity of the soil may occur due to a selective elimination of calcium cations in the exchange complex. This fact affects the availability of other chemical elements, giving rise to the fixation of phosphorus and producing free aluminum, which ultimately causes severe toxic effects in the soil.
- Reduction of the soil microbial population, as well as its microflora, which affects nitrification processes.

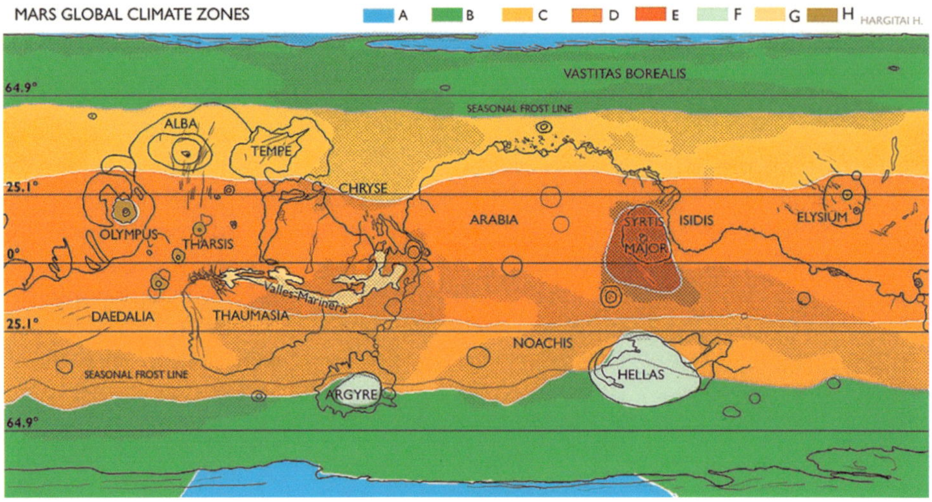

Figure 6.
Mars global climate zones [18], based on temperature, modified by topography, albedo, actual solar radiation. A = glacial (permanent ice cap); B = polar (covered by frost during the winter which sublimates during the summer); C = north (mild) transitional (C_a) and C south (extreme) transitional (C_b); D = tropical; E = low albedo tropical; F = subpolar lowland (basins); G = tropical lowland (chasmata); H = subtropical highland (mountain).

It is evident, after analyzing the shape and dimensions of the cross-section profiles made in the study area, that the main channel has the typical shape of a gully. In a broad sense, a gully is a deep depression, channel or ravine, very active for the natural drainage of a liquid fluid that flows through the surface. Although gullies can be continuous and discontinuous, it should be taken into account that the last one appears when the gully bed has a lower slope level than the general slope of the surrounding terrain.

On the other hand, it is also known that, in a gully, runoff is channeled into trenches that deepen over time to form a marked front (head) with very steep faces (walls). The gullies extend and deepen in an upward direction due to cascading erosion and progressive collapses of their upper parts. Similarly, gully slopes can collapse due to seepage of water, as well as undermining the flow of water within the gully.

With regard to the Earth (**Figure 7**), it should be noted that gullies tend to form where the slopes are long and there has been a logical loss of vegetation. This fact has as a consequence an increase in surface runoff. In particular, they tend to be dominant in areas where the soil is composed of deep silty or clay materials, in soils with unstable clays (as in the case of sodium soils), under bare rocky surfaces and on very steep slopes subject to water infiltration and earthmoving.

In another vein, and in relation to the cross-sectional profiles obtained, **Table 1** shows the area of each one of them. As can be seen, from cross-section profile number 8 (km 347 from start in **Figure 8**), inclusive, there is a significant increase in the surface of the gully, perhaps due to the contribution made by an effluent or another contributing watershed, such as [19] specifies.

Regarding the volume of soil lost (Eq. 1) in the study area, it should be noted that a result of 10,817.812 km^3 has been obtained. If this result is compared with that obtained by [19–21] (**Table 2**), it can be seen that the methodology used affects the calculation of the volume, so much so that, if Eq. (3) would have been used

Figure 7.
Gullies of Cerro Negro (Castilla La Mancha, Spain).

Profile no.	Top edge width (km)	Base width (km)	Depth (km)	Area (km^2)
1	11.410	2.992	0.677	4.87
2	9.342	2.215	0.710	4.10
3	12.042	3.078	0.545	4.12
4	15.820	1.356	0.550	4.72
5	15.077	5.786	0.590	6.15
6	12.139	2.388	0.607	4.41
7	10.540	3.260	0.651	4.49
8	27.185	11.273	1.393	26.78
9	35.925	20.118	0.649	18.17
10	18.865	9.894	1.657	23.83
11	18.056	6.216	1.731	21.01
12	18.615	4.875	1.820	21.37
13	18.409	7.699	1.648	21.51
14	17.850	9.135	1.390	18.75
15	25.193	8.541	1.632	27.52
16	30.510	1.695	1.828	29.44

Table 1.
Area in each cross-sectional profiles.

(**Figure 9**), the soil lost volume, in the present work, would have been 11,553.29 km^3. Calculation of this volume using the Topocal 2022-v9.0.811 software resulted in 12,700 km^3.

$$V\left(km^3\right) = \frac{d}{6}(A_1 + 4 \times A_m + A_2) \tag{3}$$

Regarding the RCI shown in Eq. (2), each of its variables has been obtained as follows:

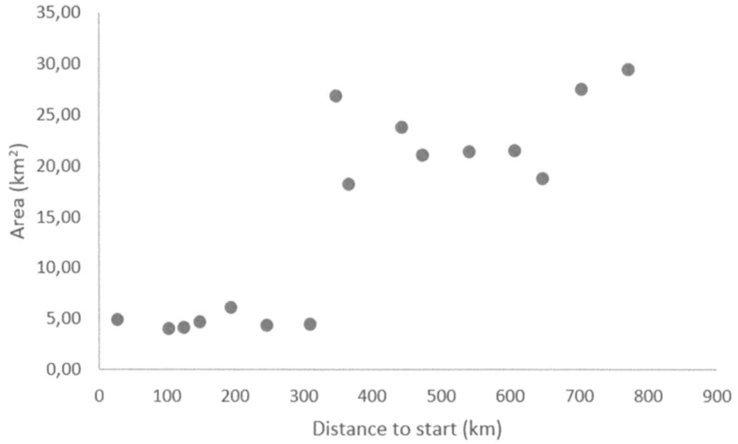

Figure 8.
Relationship between the area of each cross-section profile and its distance to start.

Author or work	Methodology	V (km^3)
Cabrol et al. (1996) [19]	Geometric model	13,900
Goldspiel and Squyres (1991) [20]	Contributing watersheds	13,000
Irwin III et al. (2002) [21]	Sedimentary transport	15,100
This work	Eq. (1)	10,817.812
This work	Prismatoid equation (Eq. 3)	11,553.29
This work	Topocal 2022-v9.0.811	12,700

Table 2.
Soil lost volume in Ma'adim Vallis by different methodologies.

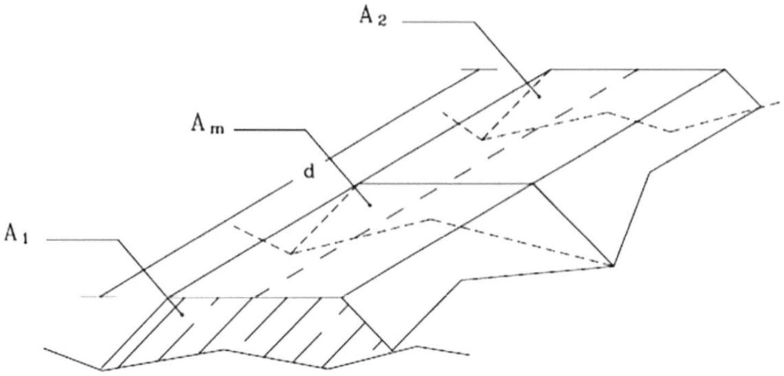

Figure 9.
Sketch to obtain the volume by the prismatoid equation. A_m *is the mean cross-section area between* A_1 *and* A_2.

The Q'_s value, dependent on the mass ("m" in g) of soil lost and of the contributing watershed area ("A" in m^2). According to [22], the average density of soil on Mars is 2.58 g/cm^3. Taking into account, as soil lost volume, for having been calculated according to the MOLA data, that corresponding to that obtained through Topocal 2022-v9.0.811, that is, 12,700 km^3, the soil mass amounts to the value of:

$$m = 12700\ \text{km}^3 \times 2.58\frac{\text{g}}{\text{cm}^3} \times \frac{10^{15}\ \text{cm}^3}{1\ \text{km}^3} = 3.28 \times 10^{19}\ \text{g} \tag{4}$$

According to [16], the contributing watershed area is 208,500 km^2 (2.085 10^{11} m^2), so the Q'_s value is:

$$Q'_s = \frac{3.28 \times 10^{19}\ \text{g}}{2.085 \times 10^{11}\ \text{m}^2} = 1.57 \times 10^8 \frac{\text{g}}{\text{m}^2} \tag{5}$$

If the soil lost volume is converted to an equivalent per km^2, it is obtained (12,700 km^3/208,500 km^2) 0.0609 km^3/km^2.

In order to obtain the average elevation ("H" in m) of the study area, data shown in **Figure 3** were taken into account. It should be noted that the average elevation does not refer to that calculated with respect to level 0, since, in this case, the result varies depending on the absolute elevation at which the study area is located, which does not would make sense. For this reason, the average elevation is obtained taking into account that level 0 is the lower level of Ma'adim Vallis, so H will be the average elevation at said level, that is, 2 km (431 m − 1569 m).

In order to infer the average slope ("S" in °) of the contributing watershed, and knowing that the gully length is 771 km, it will only be necessary to perform the following calculation:

$$S = \tan^{-1}\left(\frac{2\ \text{km}}{771\ \text{km}}\right) = 0.149° \tag{6}$$

Finally, substituting all the numerical values in Eq. (2), the following valid expression for Ma'adim Vallis is obtained:

$$\text{RCI} = e^{\frac{1.56 + \log\left(\frac{1.57\times10^8}{t'}\right) - 0.46\times\log(2000)\times\tan(0.149)}{2.65}} = e^{\frac{1.556 + \log\left(\frac{1.57\times10^8}{t'}\right)}{2.65}} \tag{7}$$

It is necessary to emphasize that the developed model has been calculated taking into account the runoff of liquid water due to rainfall in the form of rain that, in the past, would have occurred on Mars. In addition, as is well known, on the red planet the sublimation of groundwater, in low pressure atmospheric conditions, can cause a landslide analogous to that produced by runoff of liquid water flowing on the surface, a phenomenon that should be taken into account in future studies in order to be able to obtain all the causes that can affect the soil loss on Mars.

5. Conclusions

In present work, the RCI has been obtained assuming that, exclusively, similarly to what happens on Earth, the soil loss on surface is caused mainly by the action of runoff water that circulates down slope.

It is evident that the collected data to date, in relation to the relationship between erosion and rainfall, are insufficient to be able to reach a reliable conclusion. However, a possible more global development, of the presented RCI, valid for all planets, may be important when deciding the fate of future missions to Mars, especially with regard to the study of the action of water on Martian landscape.

Studies on this field have been carried out on Earth, which is why it will be necessary to start from these in order to develop algorithms, or mathematical models, capable of predicting, or calculating, the action of water globally (interaction between soil, possible existence of organisms on the surface and atmosphere).

For this reason, it is evident that future missions to Mars must be developed under the supervision of a multidisciplinary team of experts in all areas of knowledge, that is, biologists, physicists, chemists, astrophysicists, geologists, computer scientists, aerospace engineers and agronomical engineers among others.

Acknowledgements

This work has been possible thanks to the use of equipment from the University of Seville, as well as the Technology-Based Company RS3 Remote Sensing SL.

Conflict of interest

The author declares no conflict of interest.

Author details

Emilio Ramírez-Juidías
Graphic Engineering Department, University of Seville, Seville, Spain

*Address all correspondence to: erjuidias@us.es

References

[1] Manske L, Marchi S, Plesa A-C, Wünnemann K. Impact melting upon basin formation on early Mars. Icarus. 2021;**357**:114128. DOI: 10.1016/j.icarus.2020.114128

[2] Barnes JJ, McCubbin FM, Santos AR, Day JMD, Boyce JW, Schwenzer SP, et al. Multiple early-formed water reservoirs in the interior of Mars. Nature Geoscience. 2020;**13**:260-264. DOI: 10.1038/s41561-020-0552-y

[3] Lammer H, Scherf M, Kurokawa H, Ueno Y, Burger C, Maindl T, et al. Loss and fractionation of noble gas isotopes and moderately volatile elements from planetary embryos and early Venus, Earth and Mars. Space Science Reviews. 2020;**216**(4):1-65. DOI: 10.1007/s11214-020-00701-x

[4] Brasser R, Mojzsis SJ, Matsumura S, Ida S. The cool and distant formation of Mars. Earth and Planetary Science Letters. 2017;**468**:85-93. DOI: 10.1016/j.epsl.2017.04.005

[5] Ramírez-Juidías E, Villavicencio-Valero K, Borja A. Sedimentation and proposed algorithms to detect the possible existence of vegetation and humidity in the landing area of the Mars Exploration Rover-B (Opportunity). In: Bevelacqua J, editor. Solar System Planets and Exoplanets. London, UK: IntechOpen; 2021. pp. 77-95. DOI: 10.5772/intechopen.97628

[6] Mojzsis SJ, Brasser R, Kelly NM, Abramov O, Werner SC. Onset of gigant planet migration before 4480 million years ago. The Astronomical Journal. 2019;**881**(44):1-13. DOI: 10.3847/1538-4357/ab2c03

[7] Andrews-Hanna JC, Zuber MT, Banerdt WB. The Borealis basin and the origin of the martian crustal dichotomy. Nature. 2008;**453**:1212-1215. DOI: 10.1038/nature07011

[8] Ehlmann BL, Edwards CS. Mineralogy of the Martian surface. Annual Review of Earth and Planetary Sciences. 2014;**42**:291-315. DOI: 10.1146/annurev-earth-060313-055024

[9] Livengood TA, Kostiuk T, Hewagama T, Smith RL, Fast KE, Annen JN, et al. Evidence for diurnally varying enrichment of heavy oxygen in Mars atmosphere. Icarus. 2020;**335**:113387. DOI: 10.1016/j.icarus.2019.113387

[10] Kass DM, Schofield JT, Kleinböhl A, McCleese DJ, Heavens NG. Mars climate sounder observation of Mars' 2018 global dust storm. Geophysical Research Letters. 2020;**47**(23):1-9. DOI: 10.1029/2019GL083931

[11] Hockey T, Bartlett J, Boice D. Rusty Mars. In: Hockey T, Bartlett JL, Boice DC, editors. Solar System. Boca Raton: CRC Press (Taylor & Francis Group); 2021. pp. 101-114. DOI: 10.1201/9781003197553

[12] Craddock RA, Lorenz RD. The changing nature of rainfall during the early history of Mars. Icarus. 2017;**293**:172-179. DOI: 10.1016/j.icarus.2017.04.013

[13] Raack J, Conway SJ, Herny C, Balme MR, Carpy S, Patel MR. Water induced sediment levitation enhances downslope transport on Mars. Nature Communications. 2017;**8**:1151. DOI: 10.1038/s41467-017-01213-z

[14] Seybold H.J., Kite E., Kirchner J.W. The Role of Surface Water in the Geometry of Mars' Valley Networks and Its Climatic Implications [Internet]. 2017. Available from: https://arxiv.org/abs/1709.09834v1 [Accessed: 01-12-2021]

[15] Leverington DW. Incision of Ma'adim Vallis (Mars) by dry volcanic magafloods effused from multiple highland source. Planetary and Space Science. 2020;**191**:105021. DOI: 10.1016/j.pss.2020.105021

[16] Irwin RP III, Howard AD, Maxwell TA. Geomorphology of Ma'adim Vallis, Mars, and associated paleolake basins. Journal of Geophysical Research. 2004;**109**(E12):1-33. DOI: 10.1029/2004JE002287

[17] Fournier F. Climat et erosion: La relation entre l'érosion du sol par l'eau et les precipitations atmosphériques. 1re éd. Paris: Presses Universitaires de France; 1960. p. 206. https://horizon.documentation.ird.fr/exl-doc/pleins_textes/2021-11/010080384.pdf

[18] Hargitai H. Mars Climate Zone Maps Based on TES Data [Internet]. 2009. Available from: http://planetologia.elte.hu/mcdd/climatemaps.html [Accessed: 15-12-2021]

[19] Cabrol NA, Grin EA, Dawidowicz G. Ma'adim vallis revisited through new topographic data evidence for an ancient intravalley lake. Icarus. 1996; **123**:269-283. DOI: 10.1006/icar.1996.0157

[20] Goldspiel JM, Squyres SW. Ancient aqueous sedimentation on Mars. Icarus. 1991;**89**(2):392-410. DOI: 10.1016/0019-1035(91)90186-W

[21] Irwin RP III, Maxwell TA, Howard AD, Craddock RA, Leverington DW. A large paleolake basin at the head of Ma'adim Vallis. Mars. Science. 2002;**296**:2209-2212. DOI: 10.1126/science.1071143

[22] Goossens S, Sabaka TJ, Genova A, Mazarico E, Nicholas JB, Neumann GA. Evidence for a low bulk crustal density for Mars from gravity and topography. Geophysical Research Letters. 2017; **44**(15):7686-7694. DOI: 10.1002/2017GL074172

Chapter 6

Impacts of the Industrial and Technological Revolution on Territories and Cities in the Arid Environments: A Case Study of the Antofagasta Region in the Atacama Desert, Chile

José Norberto Guerra Ramírez

Abstract

This chapter addresses the environmental and socio-cultural impact that the Industrial Revolution has maintained in the exploitation of the arid territories of the Atacama Desert. The perspectives, challenges and management of a region noted for its great contribution to the economic development of Chile are observed. Since the first promotion of the mining industry, today new industries such as solar energy, tourism and astronomy have strengthened regional development. This is the region with the highest solar radiation on the planet and one of the best places for astronomical observation, in the most Mars-like landscape on Earth. However, the development has not manifested itself in the well being of the cities, which have an average standard of urban development. The methodology of the study has a qualitative approach, assessing the fragility of the desert ecological system, with the strategies of adaptation to climate change and sustainable development. Improving the livability of cities, with unprecedented urban and architectural forms, is a necessity. The research hypothesis proposes that desert territories have a great potential to trigger sustainable energy transition. In conclusion, deserts are key environments to lead the process of innovation and creation of new desert cities.

Keywords: new technologies, climate change adaptation, cities in the Antofagasta region, Atacama Desert

1. Introduction

The energy industrial revolution in the Atacama Desert has a relevant milestone with the inauguration in June 2021 of Cerro Dominador, the first solar thermal plant in

Latin America. This circumstance documented by the report by the journalist Nick Miroff of The Washington Post, on March 31, 2017, allows us to dimension the value of solar energy for Chile, classified as the "Solar Saudi Arabia" [1].

The Atacama Desert is located in northern Chile, administratively between the Antofagasta Region to the north and the Atacama Region to the south, with an extension of more than 1000 km between latitudes 19°S and 30°S. As indicated by researcher Rodrigo Palma, the potential of the Atacama Desert to generate electricity thanks to the high solar radiation could supply 60 times the consumption of Chile and 20% of the world" (**Figures 1** and **2**) [2].

The altitudinal component determines the different environments and characteristics of aridity according to altitude and distance from the sea, three ecological floors can be distinguished, such as: the coastal desert, strip of territory between the marine plains and the Cordillera de la Costa, this area with oceanic influence and the cold Humboldt current, leave a region with high atmospheric humidity and saline coastal environment. The intermediate desert includes the intermediate depression characterized by low relative humidity and clear skies, recognized as the ecological floor of the pampas, it is the driest and most arid area of the planet. It is the driest and most arid zone on the planet. Here there is an important rain shadow, which prevents the advection of humidity to a great extent. The interior deserts, considered the pre-mountain range and high plateau of the Andes Mountains, in this area are the first vestiges, a culture of hunter-gatherers, the natural shelter of the ravines gives rise to human settlements and microclimatic oasis, ideal for subsistence agriculture and livestock.

Technological innovation and the industrial revolution have been fundamental to create a productive infrastructure that financed road works and public facilities, favoring the growth of urban centers [3]. The connection between mining and industry in the Atacama Desert has been key to build the habitability of mining settlements and promotes the development of these extremely arid territories, with an

Figure 1.
Cerro Dominador: The first Concentrated Solar Power Plant in Latin America with 10,600 mirrors (heliostats) of 140 m^2 each and more than 1,000 hectares, it is capable of avoiding the emission of about 640,000 tons of CO_2 per year and supplying a city of approximately 380,000 homes, located near the saltpeter office of Maria Elena in the Atacama Desert, Chile.

Figure 2.
Cerro Dominador solar platform view in Atacama Desert landscape.

architecture that integrates passive elements of environmental conditioning and prefabrication and modulation technology.

The exploitation of natural resources and their management allowed the growth of port cities, the exploitation of saltpeter between 1880 and 1930 allowed the occupation and urbanization of the central strip or intermediate plain of the Atacama Desert, it is here, in the middle of the saltpeter pampas, in extremely arid and dry environmental conditions, where the "saltpeter cities" are located (**Figure 3**). The railroad industry with the first steam trains allowed the penetration and conquest of these new uninhabited and unexplored landscapes, incorporating them to this day to the productive map of the region, connecting the industrial enclaves from the coast of Antofagasta with the desert of the altiplano to Uyuni in Bolivia [4]. There were 118 saltpeter offices, in which more than 46,470 workers worked, this generated an unprecedented population of men and women, developing their daily lives and cultural development in these inhospitable environments, mining settlements, and workers who knew how to adapt to the working conditions of great physical effort of the mining work and the environmental conditions of high solar radiation, low humidity, water shortages, and high-temperature differences day-night, with all this created a culture and society with heterogeneous and complex pampino identity. Its unity is given by the shared space: the pampa calichera in the middle of the driest desert in the world, and by the work, linked in all its aspects to the nitrate mining operations.

Technological innovation and the industrial revolution have been fundamental to create a productive infrastructure that financed road works and public facilities, favoring the growth of urban centers [3]. The connection between mining and industry in the Atacama Desert has been key to build the habitability of mining settlements

Figure 3.
Desert Cities: Arturo Prat Saltpeter Office, in abandoned ruins in the "Cantón Central" sector of the Atacama Desert, Sierra Gorda.

and promotes the development of these extremely arid territories, with an architecture that integrates passive elements of environmental conditioning and prefabrication and modulation technology.

The exploitation of natural resources and their management allowed the growth of port cities, the exploitation of saltpeter between 1880 and 1930 allowed the occupation and urbanization of the central strip or intermediate plain of the Atacama Desert, it is here, in the middle of the saltpeter pampas, in extremely arid and dry environmental conditions, where the "saltpeter cities" are located (**Figure 3**). The railway industry with the first steam trains allowed the penetration and conquest of these new uninhabited and unexplored landscapes, incorporating them until today to the productive map of the region, connecting the industrial enclaves from the coast of Antofagasta with the desert of the altiplano to Uyuni in Bolivia [4]. There were 118 saltpeter offices, where more than 46,470 workers were employed, which generated an unprecedented population of men and women, developing their daily lives and cultural development in these inhospitable environments, mining settlements, and workers who knew how to adapt to the labor conditions of great physical effort of mining work and the environmental conditions of high solar radiation, low humidity, water shortages, and high-temperature differences day-night, with all this created a culture and society with heterogeneous and complex pampina identity. Its unity is given by the shared space: the pampa calichera in the middle of the driest desert in the world, and by work, linked in all its aspects to nitrate mining (**Figures 4** and **5**).

Undoubtedly, the copper industry has been fundamental in projecting perspectives, in defining the challenges and management processes of the Atacama Desert, driving territorial transformations, the natural landscape, infrastructure, and quality of life in the cities, but it has also been a source of environmental conflicts associated with the use of water that has affected the flora and fauna of the salt flats and the subsistence agricultural economies of the region [5]. Andean communities. However, a challenge for the development of the region is represented by its mono-productive and exporting characteristics of nonrenewable natural resources, basically minerals, which, as we have already mentioned, have a great impact on the national budget and development. The new industries in development today are changing this condition opening to sustainable development, therefore, solar energy and lithium represent a new source of economic, productive, and social development, they are a new future that allows expanding the productive diversification of these arid territories, responding in a better way to the objectives of sustainable development of the UN [6].

In 1998, the lithium industry placed Chile in the first place as a world producer with known reserves in the South Atacama of about 40% of lithium in brines [7]. In the current context of energy transition toward green and clean energies, the demand for electromobility based on lithium batteries has become increasingly relevant, which represents a future development opportunity for the region, however, the biggest challenge is how to obtain lithium in an environmentally responsible manner, without devastating the fragility of the ecosystem of the Salar de Atacama basin, the destruction of its water sources, its geological, ecological and cultural heritage of the driest desert in the world (**Figures 6** and 7) [8].

Figure 4.
Chuquicamata Mine, Calama Chile.

Figure 5.
Trucks at work in the Chuquicamata mine.

In extreme arid environments, inventiveness and creativity have been put to the test for the adaptation and development of new technologies, for example: The first solar desalination plant in the Domeyko Office, Canton el Boquete, near the port city of Taltal, in the Atacama Desert, built in 1906–1908, supplied the vital element to the population of the saltpeter camp (**Figure 8**) [9, 10].

How to obtain fresh water in a desert environment with no rainfall and high solar radiation for the needs of the saltpeter industry was the challenge that the Compañía Salitre de Boquete, Antofagasta, Chile, set out to solve, for which they designed an apparatus composed of a series of frames containing 1850 m^2 of glass. The panels are arranged in a V-shape, and under each panel is a shallow tray containing brackish water. The heat from the sun condenses the water on the sloping glass, and this freshwater flows down into a small channel and is carried into the main channel. A total of 3600 liters of fresh water can be collected daily [11].

Another of these innovations are the "fog catchers" for obtaining water in certain areas of the coastal desert. These are cloud capture systems formed by meshes that intercept the fog and, through a system of gutters, collect the water trapped by the mesh. This has allowed the development of fog oasis agriculture associated with the

Figure 6.
Lithium mining in the Salar de Atacama contrasts with the fragility of the ecological.

Figure 7.
Hydric ecosystem of the Salar de Atacama basin.

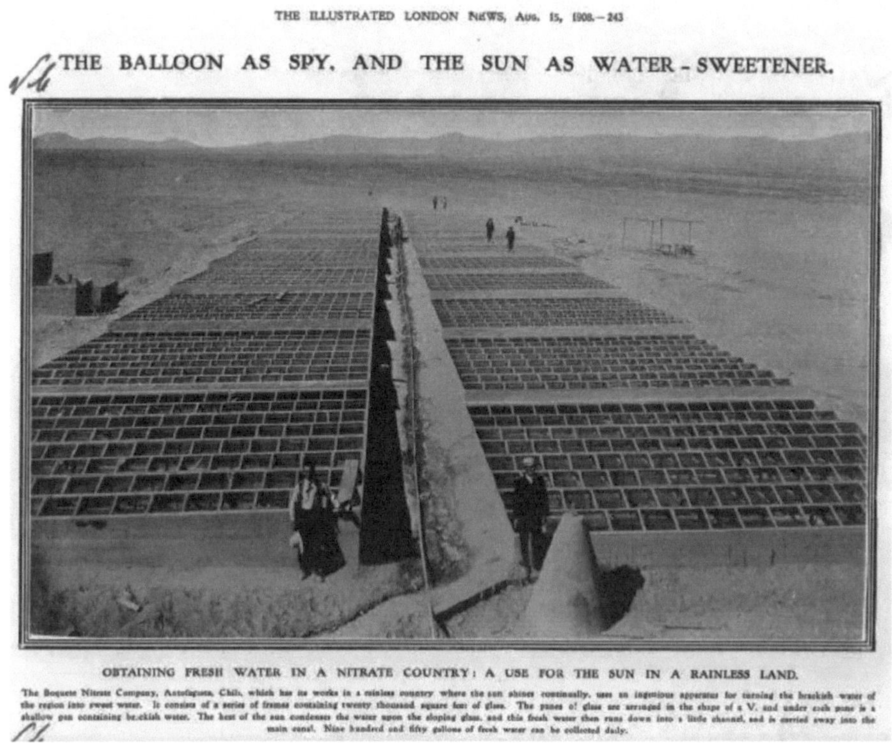
THE ILLUSTRATED LONDON NEWS, AUG. 15, 1908.—243

THE BALLOON AS SPY, AND THE SUN AS WATER-SWEETENER.

OBTAINING FRESH WATER IN A NITRATE COUNTRY: A USE FOR THE SUN IN A RAINLESS LAND.

The Boquete Nitrate Company, Antofagasta, Chili, which has its works in a rainless country where the sun shines continually, uses an ingenious apparatus for turning the brackish water of the region into sweet water. It consists of a series of frames containing twenty thousand square feet of glass. The panes of glass are arranged in the shape of a V, and under each pane is a shallow pan containing brackish water. The heat of the sun condenses the water upon the sloping glass, and this fresh water then runs down into a little channel, and is carried away into the main canal. Nine hundred and fifty gallons of fresh water can be collected daily.

Figure 8.
Solar desalination plant of Oficina Domeyko, Cantón el Boquete Atacama Desert, built in 1906–1908. Photograph retrieved by Maria Telkes. Available in The Illustrated London News and other newspapers in England and the United States. las-salinas-copia1.jpg (414 × 234) (wordpress.com).

various passive bioclimatic strategies of environmental conditioning, ventilation, and natural heating make it possible to live in extreme climatic conditions (**Figures 9–12**).

According to Cereceda's study on the origin and behavior of fog in the Coastal Cordillera and inland localities of Atacama, it was found that fogs move inland toward the continent with different intensities depending on the season and time of day, and the maximum spatial extension occurred during winter and night, up to 12 km inland [12].

In the same line of the use of renewable resources, which allow the transformation and decarbonization toward a clean energy matrix, wind farm projects continue to be developed in the Atacama Desert (**Figures 13** and **14**) the next one called Horizonte of the company Colbún will be the largest in Latin America, which contemplates the installation of 140 wind turbines, which will allow an average annual generation of 2,400 Gwh, equivalent to the energy consumption of 700,000 homes. The implementation of the project will avoid the emission of 1.2 million tons of $CO2$ per year [13].

On the other hand, the Atacama Desert concentrates most of the world's astronomical observation, with two important scientific observation centers, the Cerro Paranal observatory in operation since 1996 (**Figures 15** and **16**) and the observatory at Cerro Armazones, which is under construction and will be operational in 2027. What makes this desert special is a combination of geographic and climatic factors. With an altitude of about 3,000 meters above sea level near the coast, an extremely dry environment, very low probability of rain, little water vapor in the air, an average

Figure 9.
Bank of clouds developed between 300 and 1200 meters above sea level, in the coastal mountain range, generate the "oasis of fog". In this zone, cool, moisture-laden air is trapped under a layer of warm air above.

Figure 10.
Original structure of the macro diamond fog catcher installed in the "Los Nidos" sector of the Cordillera de la Costa, in the image its inventor Dr. Carlos Espinosa Arancibia (1924–2022) being interviewed in situ by French TV for the international exhibition "Le bord des mondes" held in Paris in 1995.

wind speed of 25 km/h, and almost no light pollution, the Atacama Desert is the best viewpoint to the galaxies and allows the promotion of new productive activities for these desert territories of Atacama [13].

Figure 11.
Structure inspired by the "Atrapanieblas" corresponds to the 2011 Integrated Exercise dedicated to the design of an ecology for the coastal desert of Atacama carried out by the basic cycle workshops (Shelter, Resource, Language and Context workshops) of the School of Architecture of the Universidad Católica del Norte, Antofagasta, Chile.

Figure 12.
Structure inspired by the "Atrapanieblas".

Figure 13.
Calama and Valle de los Vientos wind farms near the city of Calama, on the RCH 23 highway that connects to San Pedro de Atacama.

Figure 14.
Wind mills in the surroundings of the city of Calama.

Figure 15.
Cerro Paranal Observatory.

Figure 16.
Hotel for Astronomers at 2635 meters above sea level in the Cordillera de la Costa.

Figure 17.
Solar photovoltaic plant Uribe Solar Station near the city of Antofagasta.

The Antofagasta region in northern Chile has the highest solar radiation on the planet. Plant factors of 35% are achieved in single-axis tracking PV plants. The global average plant factor for solar PV plants was 11% in 2020. In the United States, the average was 20% between 2018 and 2020 [14].

This condition has allowed the rapid growth of the solar industry, in 6 years, Chile has quintupled its generation capacity from renewable energy sources and it is projected that, by 2030, up to 70% of its electricity matrix will be clean. The growing investment in these energies, as well as in storage and transmission infrastructure, is a clear sign of a determined transition toward a more sustainable electricity system that shows that the future prospects of the Atacama Desert are very promising for the region (**Figures 17** and **18**).

The development model of the Antofagasta Region based on the mining industry has followed a pattern of growth, making important contributions to the national GDP with a low level of local development, as evidenced by the low quality of its urban infrastructure. Currently, there has been a high investment in the solar industry that has put the region and cities in northern Chile such as Arica, Iquique, Antofagasta,

Figure 18.
The solar photovoltaic plant designed has a maximum power of 56,632,400 Wp consisting of 182,960 solar modules of 315 Wp, 336 solar trackers on an east-west axis.

Calama, and Copiapó back on the world stage, projecting new impulses and challenges of productive development for the entire macro-region of the Atacama Desert [15].

The city of Antofagasta is the mining capital of Chile, with a population of approximately 400,000 inhabitants. Antofagasta's GDP per capita is approximately US$40,000 (nominal in 2017), a figure equivalent to that of countries such as South Korea or New Zealand. However, this is not reflected in the urban and building quality of the city of Antofagasta, this being a reality that applies to all cities in northern Chile [16].

The following perspectives, challenges, and management can be derived from this:

1.1 Perspectives

Are to scale the results and contributions of the region to improve the quality and identity to improve the quality of the infrastructure of the northern cities. For this, it is key to learn from the past by taking advantage of specific design solutions for the local micro-climatic determinant.

1.2 Challenges

Are to continue contributing to the country, but with greater investment and environmental responsibility, to design cities in harmony with the desert ecology and adaptation to climate change. This challenge implies a great capacity for co-creation, since unprecedented proposals must be generated that integrate the views of all relevant actors in the city to turn it into an excellent model and example of development on a human scale.

1.3 Management

It is to design cities that maintain their human scale, more egalitarian, with less inequality and greater identity, with public spaces and buildings that can meet the needs of healthy and safe habitability for people and with technological conditions so that through data intelligent decisions can be made.

2. Perspectives, challenges, and management

2.1 The perspectives

As detailed above, the contribution of arid environments on a productive scale is very significant and even more so in a context of climate change and energy transition.

The perspectives are to scale the results and contributions of the region to improve the quality and identity to improve the quality of the infrastructure of the northern cities. To this end, learning from the past by leveraging design solutions specific to the local microclimate determinant is key [17].

The disconnect in urban and building development is not new and derives from the central vision, and the absence of adequate infrastructure, which is also based on central decisions that are not very conscious and respectful of local realities and needs and end up in unfortunate experiences. The urban development of the city of Antofagasta is very abandoned, planning arrives late and is not articulated with the growth of the city and this generates disorder, generates spontaneous occupation that requires long and costly regularization processes to be in accordance with the regulations in force (**Figure 19**).

There are undoubtedly notable exceptions in architecture that also allow us to confirm this situation. In the case of Antofagasta, there are buildings such as the "Colectivos obreros," the Caliche building, and the Curvo building that do represent a modern architecture that adapts and brings together elements of languages very relevant to the environmental conditions. Issues such as fixed or movable eaves to shade the window and the interior of the room are repeated very efficiently in these buildings.

If the construction of urban shade and shadow in the envelope is an invariant element that is recognized in vernacular and contemporary architecture. At the urban and building level, it is required that the cities of arid environments develop proposals unpublished to their environment, that the proposals are correct to their environmental determinant, and take advantage of passive design solutions to develop proposals adapted to the climate, here the fundamental element that takes all the protagonist, is the appropriate response to solar orientation. Depending on the form and architectural program, the provision to solar gain from the north should be considered adequately because the living spaces to the south are often very uncomfortable.

There is an excellent opportunity to generate an architectural and urban language that innovates by being faithful to passive environmental conditioning strategies applied to architectural envelopes and new materials. The double skin, double roof, ventilated facade, and cross ventilation in the coastal desert and the pampas desert are very relevant strategies. In addition, vine-type plant facades provide moisture and color to facades in arid climates (**Figures 20–25**).

Shade is an architectural element that contributes to the identity of the urban space of desert cities, it must be present because it is a health requirement, given the high rates of skin cancer, for the welfare and enjoyment of public space. With this we will be promoting the importance of the square meters of shade in the public space of desert cities, in favor of this indicator itself and not of the square meters of green areas that will always be far below what the World Health Organization indicator establishes. Shaded structures on an urban scale should have as design premises the

Figure 19.
Self-construction and informal settlements that defy construction and safety standards.

Figure 20.
Case of a house in the Chacabuco saltpeter office, with a shaded corridor around the perimeter and a double ventilated roof and shading system.

Figure 21.
Case of a Chalet house in the María Elena saltpeter office, with a ventilated façade with wooden trellised treillage so that the vine or climbing vegetation covers the architectural envelope.

Figure 22.
Façade with fixed shade house vernacular house of María Elena.

Figure 23.
Contemporary façade of new building Municipality of María Elena.

Figure 24.
Capture of a woman in the streets of Antofagasta in search of sun protection and Fig.

Figure 25.
Shaded shelter in the Trocadero Beach Resort, with great performance in terms of its design, providing continuous shade for the enjoyment of the sheltered walk.

continuity of shade, a height, width, and filter quality that does not generate opaque or hard shade, but lets light filter through in a subdued manner. These are solutions that were very present in the vernacular architecture of the northern cities and it is very necessary to recover from the past to put them in the present (**Figures 24** and 25).

At this point, it must also be said that the most suitable materials are not plastic sheds because they generate heat and uncomfortable brightness, the best are the natural elements that also have good behavior to the stress of high saline humidity of the coastal desert and resistance to high radiation and dryness of the intermediate desert.

2.2 The challenges

When the city of Antofagasta has been affected by intense rainfall events in the present decade, when these phenomena were of very low occurrence and that this is associated with a human tragedy claiming lives, it is logical to think that if the creeks were dry for years and in a few hours are transformed into imposing beds of water and mud that buried parts of the city means that we have not designed cities observing the reality drawn in its geography. An urgent challenge is to plan safe cities in harmony with the ecology of the desert, it is urgent to carry out only corrective measures such as large infrastructure works to contain the force of water such as alluvial roads, but it deserves a review of the urban design of the city, planning of safe areas, where natural streams should mark and define lines of a conscious design, with areas of land use restriction for housing, assigning them to floodable park areas [18].

The urban sprawl of the city due to the explosive growth of its population due to various phenomena including high migration, construction costs, and high housing availability deficit, has given rise to illegal and uncontrolled land occupation processes, with high risk, given the occupation of the ravines at increasingly higher elevations of the Coastal Mountain Range (**Figures 26** and 27).

On the other hand, the rains also bring a completely different phenomenon to the tragedy of the floods, the water brings the awakening of the seeds that were sleeping on the surface of the desert, the magic of the blooming desert covers the always barren

Figure 26.
Occupation of informal and self-built housing on the hills and ravines of the City of Antofagasta.

Figure 27.
Alluvial via in the ravines, city of Antofagasta.

Figure 28.
Desert in bloom on hillsides in the city of Antofagasta.

desert slopes, for a few days the magic happens and the desert landscape is transformed into a flowery spectacle (**Figures 28** and **29**).

The Atacama Desert is renowned for being the oldest and continuously driest nonpolar temperate desert on Earth, believed to have been arid since the Jurassic period and to have gradually evolved into hyperacidity during the Miocene epoch more than 5 million years ago. In addition, exceptionally high levels of ultraviolet radiation, extreme aridity, low or no soil carbon concentrations, and the presence of strong oxidizing conditions and/or toxic elements in particular habitats make the Atacama an excellent example of the extreme biosphere, the environmental limits that define life on Earth [18].

2.3 The management

According to the United Nations Intergovernmental Panel on Climate Change report, the vulnerability of the city of Antofagasta to climate hazards is very likely for components such as floods, storms, and landslides (alluvium, landslides, debris

Figure 29.
Flowerful desert, after an untraditional rainfall event.

flows) and very likely for components such as sea level rise, coastal flooding, droughts, high tides and strong waves, landslides, debris flows) and probable for components such as sea level rise, coastal edge flooding, droughts, high tides, and strong waves are high, implying a significant management of these hazards, which are part of situations that have occurred more frequently and are part of real and concrete evidence [19].

Given the conditions described above, the process of adaptation to climate change is an indicator that should be projected in urban and territorial development projects of cities in arid areas. In the particular case of the city of Antofagasta, the impact is double and complex given its location in a narrow coastal plain between the sea and the coastal mountain range, and therefore the rise in sea level and the possibility of alluvium caused by rainfall will have devastating impacts on the environment and especially on the most disadvantaged groups that are informally located on the hillsides and ravines.

In general, the conceptualization processes of the urban structure of the city must be improved in the face of the vulnerabilities described above and, in particular, the design and construction conditions of housing and buildings not designed to withstand rainfall or extreme temperatures, all of which implies changes and updates to housing regulations, which is easy to propose but difficult to implement [20].

In this order, the fragility of desert environments and territories is something that has not been sufficiently repaired, being territories rather depopulated areas, transformations and aggressions to the landscape have remained without the consequence of respect for the environment, there are many areas declared saturated with pollution, mining operations without mine closures and huge pools of accumulation of liquid industrial waste (Riles) from mining companies. Among the main impacts caused by mining effluent discharges and infiltrations are the generation of clear water and acid drainage, the presence of which can result in damage to aquatic systems and plant communities, and negative effects on surface and groundwater quality [21].

Another challenge that is directly related to the quality of life in arid cities is linked to the need to scale the technologies that are developed in the productive industries of the region and contribute to the improvement of the facilities and infrastructure of the cities, in other words, desert cities have a strong imprint of the "mining camp.". We talk about camp cities, which is associated with the processes of transhumance, high mobility, and rotation of personnel working in mining operations, who move around the territory to their mining operations and do not stay or reside in the city. This has meant that there is less rootedness, and less belonging and appreciation for the city, many professionals are rather passing through and consequently this affects the projection of living in the arid territories, with it the cultural and service proposal is less attractive, despite the best attributes of the climate. All this is counterbalanced by the high costs of housing, food, health, and education services, which are very relevant considerations for families with children. Finally, a relevant variable for desert cities is that the quality of housing is in tune with living conditions, the use of outdoor space is very important because it allows outdoor activities that in other climates are performed inside the house, therefore, the possibility of having houses with a patio, the implementation of specific passive design strategies and solutions for arid climates is not a constant and therefore should be a mandatory requirement (**Figures 30** and **31**) [22, 23].

Figure 30.
The images are proposed as the synthesis of the coexistence of an Andean world that uses the resources of the place to generate its settlements.

Figure 31.
The image of the Conchi railroad bridge that crosses the Loa River as the image of the industrialized world that with technology and innovations was conquering and domesticating the extreme arid territories just like today.

3. Conclusions

The arid regions of the Atacama Desert present very promising opportunities for future development, consistent with UN Sustainable Development Goals 11 and 7, sustainable cities and communities, with the use of renewable energy resources and higher standards of energy and water efficiency savings.

It is noteworthy how the new governance and public and private organizations in the region are attuned to the need for development with greater environmental and ecological awareness, proposing new models of development of the urban landscape of desert cities, presenting corporate social and environmental responsibility declared in their missions and institutional visions, and the demands of workers and citizens to care for and worry about improving the quality of life of the built environment, incorporating the values of regional environmental identity.

The environmental determinants of climate change, residential energy costs, environmental pollution of cities declared as saturated, desalination of sea water for human consumption and industrial water, decarbonization of the energy matrix, are very present in society, which requires a high corporate commitment, and care to protect the fragile desert ecosystem, to leave behind the look of the desert as a large landfill in which we can bury pollution or leave it in the middle of the pampas, away from the eyes of criticism, however, today the complaint of citizens and communities, increasingly empowered and critical of large investment projects require compliance with environmental regulations for the care and protection of the desert environment.

The adverse environmental impacts of mining operations include an important requirement for the care of water in an environment of extreme scarcity, which has induced companies to change their groundwater use strategies in response to conflicts generated with indigenous communities and environmental organizations that protect the flora and fauna of the salt flats.

The growing recognition of the Atacama Desert as an enormous energy reserves in natural wind and solar resources, astronomical and mineral resources, and tourism resources of spatial, patrimonial, and environmental interest positions the region as a key piece of the country's strategic development, empowering the region as a pole of frontier research in the aforementioned areas.

Author details

José Norberto Guerra Ramírez
Universidad Católica del Norte, Antofagasta, Chile

*Address all correspondence to: jguerra@ucn.cl

References

[1] Available from: https://www.washingtonpost.com/sf/world/2017/03/31/while-trump-promotes-coal-other-countries-are-turning-to-cheap-sun-power/?utm_term=.26dd1edea9d1

[2] Available from: https://acera.cl/rodrigo-palma-con-todo-el-potencial-de-energia-solar-de-chile-se-podria-abastecer-60-veces-el-consumo-del-pais-y-el-20-del-mundo/

[3] Available from: https://www.elciudadano.com/entrevistas/la-desconocida-historia-de-las-primeras-plantas-desaladoras-solares-del-mundo-en-el-desierto-de-atacama ... 1/26 2022

[4] Pizarro G, Antonio J. The Antofagasta Nitrate Company, Chile. The challenge of its business modernization and strategic innovation. Estudios atacameños. 2018;**60**:133-159

[5] Meller P, Gana J. El cobre chileno como plataforma de innovación tecnológica, Cieplan Dag Hammarskjöld N°3269, piso 3, Vitacura Santiago – Chile, Triángulo. 2015

[6] Muñoz D. El mercado del cobre a nivel mundial: evolución, riesgos, características y potencialidades futuras Revista Chilena de Ingeniaría Volumen 21 N° 2, mayo – agosto. 2013

[7] Lemunguir Sepúlveda F. El rol esencial que cumple el litio para el desarrollo de la electromovilidad SQM, La tercera. 2022

[8] Segura P. Electro movilidad, litio y Salar de Atacama: Del sartén al fuego Salar de Atacama Chile sustentable. 2019

[9] Arellano N. La ingeniería y el descarte artefactual de la desalación solar de agua: las industrias de Las Salinas, Sierra Gorda y Oficina Domeyko (1872–1907). [Versión del autor]. 2015

[10] Arellano-Escudero N. Desalación de agua, History of Technology/Historia de la Tecnología, Solar Energy /Energía Solar 5, noviembre, 2012. 2012

[11] Castro V, Aldunate C, Varela V. Ocupación humana del paisaje desértico de Atacama, Región de Antofagasta. ARQ (Santiago). 2004;**57**:14-17

[12] Cereceda Troncoso P. La distribución de la niebla en Chile, Instituto de Geografía, Pontificia Universidad Católica de Chile. Revista de Geografía Norte Grande. 1989;**16**:43-49

[13] Zamarin F. Chile construirá el parque eólico más grande de Latinoamérica - Forbes Colombia Publicado 21/09/2021. 2021

[14] Palma-Behnke R, Abarca del Río R, Agostini C, Alvear C, Amaya J, Araya P, et al. The Chilean Potential for Exporting Renewable Energy (Mitigation and Energy Working Group Report). Santiago: Conocimiento e Innovación; 2021

[15] Sustentabilidad Chile apuesta por la agricultura en el desierto para enfrentar el cambio climático. Agenda País (https://www.elmostrador.cl/autor/agendapais2030/) 17 junio, 2022

[16] Correa M, Barría C, Maluenda B. Estrategia Nacional Hidrogeno verde. Ministerio de Energía; 2021

[17] Palma R. Desafíos de la energía solar en climas desérticos y alta radiación Summit Tecnológico Minería 2018 sercchile. 2018

[18] Bull AT, Asenjo JA, Goodfellow M, Gómez-Silva B. The Atacama Desert: Technical Resources and the Growing Importance of Novel Microbial Diversity. Annual Review of Microbiology. 2016;**70**:215-234

[19] Calderón G. Potenciales impactos en las zonas extremas. Chile: Agencia de Sustentabilidad y Cambio Climático; 2019

[20] Agenda País (https://www.elmostrador.cl/autor/agendapais2030/) 17 junio, 202 Chile apuesta por la agricultura en el desierto para enfrentar el cambio climático

[21] EFE. desafío de Chile: obtener litio sin destruir los salares de Atacama – 2021. Available from: https://www.elmostrador.cl/agenda-pais/2021/01/02/el-desafio-de-obtener-litio-sin-destruir-los-salares-de-atacama/

[22] Lagos G. El Desarrollo del Litio en Chile: 1984-2012, Programa de Investigación en Economía de Minerales, Pontificia Universidad Católica de Chile, Santiago, Chile, 21 de agosto de 2012. Chile: Publicación Electrónica; 2012

[23] La vida cotidiana en la pampa salitrera (1830–1930) – Memoria Chilena, Biblioteca Nacional de Chile. Publicación Electrónica. 2022

Chapter 7

Wood Quality and Pulping Process Efficiency of Elite *Eucalyptus* spp. Clones Field-Grown under Seasonal Drought Stress

Deborah Rodrigues de Souza Santos, Camila Sarto, Rafael Fernandes dos Santos, Júlia Lôbo Ribeiro Anciotti Gil, Carlos de Melo e Silva-Neto, Regina Maria Gomes, Evandro Novaes, Carlos Roberto Sette-Junior, Mario Tomazello-Filho, Rafael Tassinari Resende and Matheus Peres Chagas

Abstract

The objective of the present study is to evaluate the wood quality of five elite *Eucalyptus* spp. clones at 4 years of age from a clonal test installed in a region of seasonal drought stress in central-western Brazil focusing on pulp production. A total of 25 trees were systematically felled and disks and logs were obtained along the trunk. Wooden disks were used for density and fiber analyses and the logs were converted into chips for application in the pulping process. For the denser genotype, clone D (*E. grandis* x *E. urophylla* x *Eucalyptus tereticornis*), a thicker cell wall associated to thinner fibers results in a negative effect on the fiber quality. In contrast, clone B (*Eucalyptus pellita* x *E. grandis*), which has relatively inferior pulping performance, displayed the lowest wood density associated to wider lumen and fibers. The best growth performances in response to acclimatization and adaptation to the site strongly influences the pulp productivity, which is identified as the parameter of greatest variance between genotypes, and highlighting clone E (*E. grandis* x *E. urophylla*).

Keywords: clonal test, kraft pulp, wood technology, fiber, density

1. Introduction

Brazil is among the world's largest pulp and paper producers and has developed studies aimed at increasing knowledge on forest-to-product raw materials in an attempt to meet the growing demands and economic interest. *Eucalyptus* trees are the most common hardwood fiber sources for chemical cellulose and paper production in Brazil, with a minimum harvesting age of approximately 4.5 years and average yield of 35 $m^3 h^{-1} yr^{-1}$ [1]. This major industrial-scale interest has led to a steady increase in the extent of *Eucalyptus* plantations, which has been moved to new frontiers such as north and central-western regions of Brazil. These new frontiers are predominantly characterized by dystrophic soils and very distinct seasonal rainfall compared with traditional regions. The climatic condition specifically for the central-western region is marked by seasonal drought stress of about 5 months [2].

It is well known that water deficit is one of the most challenging factors of our times which threatens *Eucalyptus* plantation production [3]. A slight reduction in average *Eucalyptus* yield was observed in Brazil compared to previous years as a result of advancing planted areas to regions where water deficits are more severe [4]. However, little has been explored regarding the extent of the potential impacts on the wood quality for end use. From that understanding and considering that extreme drought events will happen more frequently around the world [5], new studies aligned to increase knowledge on the changes in forestry which go beyond the influence on stand volume are intended.

Eucalyptus wood characteristics may vary substantially among species and clones [6, 7], as well as by the plantation site [8–10]. Wood density and cellular structure of xylem are highly associated when considering the wood characteristics related with adaptation mechanisms to drought [11, 12]. There is a close relationship between density and hydraulic safety in the sense that a greater resistance to the cavitation process (better hydraulic safety) is related to higher wood density [13]. This in turn could be explained by changes in the structure of vessel elements [13], and/or by the contribution that the fiber matrix (non-conductive elements) makes to adjust the necessary equilibrium during water transport under high tension [14]. Hence, since the wood quality for pulp production is strongly influenced by anatomy and density properties [15, 16], the extent to which these adaptive responses determine pulp and papermaking performance has economic relevance [17].

The *Eucalyptus grandis* x *Eucalyptus urophylla* hybrid has been widely used in Brazil, providing most of the raw material which is used for hardwood pulp companies [18]. Additionally, interspecific hybrids from species such as *E. grandis*, *E. urophyla*, *Eucalyptus camaldulensis*, *Eucalyptus pellita* and *Eucalyptus tereticornis* have been utilized as a strategy for dry environments [19]. Researchers in a public-private partnership have installed an unprecedented breeding program with several *Eucalyptus* species and hybrids in Goiás state, in the central-western region of Brazil. The objective has been to evaluate the performance of different genetic materials against specific climate and soil conditions, aiming to select superior clones to compose future plantations on an industrial scale in these new *Eucalytpus* plantation frontiers in the country. Therefore, the present work evaluated five elite *Eucalyptus* clones at four-year-old grown in Goiás state, Brazil, regarding their wood properties and their impact on morphological quality indices and pulping process efficiency. The overall objective was to indicate the pulpwood potential of superior genetic materials adapted to the Brazil central-western region, contributing to increase knowledge of drought-tolerant genetic materials available for the pulp and paper industry.

2. Materials and methods

2.1 Site and wood sampling

Five elite *Eucalyptus* spp. clones at 4 years old were selected from a clonal trial located in Goiás State, central-western Brazil (16° 16′ 37″ S; 47° 44′ 02″ W; 930 m

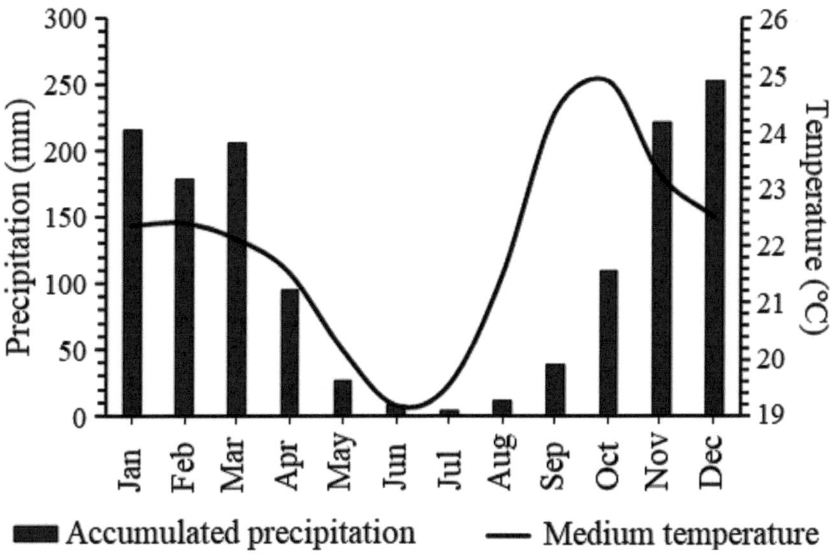

Figure 1.
Climatological normal (1990–2020) of temperature (°C) and accumulated precipitation (mm) of Luziânia, Goiás, Brazil - NASA power climate data.

above sea level). The experiment was established in December 2012, with 93 *Eucalyptus* species/hybrids, using single tree plots and 29 replications. Trees were planted at 3.0 × 3.0 m spacing and were managed according to the operational practices for *Eucalyptus* plantations in Brazil (i.e., with limestone and NPK fertilization, and control of weeds and ants). According to the Köppen-Geiger climate classification, the region's climate is subtropical humid with dry winters and wet summers [20]. The mean annual rainfall is near 1300 mm with five dry months per year, and the annual average temperature is 22.0°C (**Figure 1**). The soil is Plinthosol, being characterized as gravelly with an acidic pH, low base saturation and high aluminum saturation.

A total of 25 trees were systematically felled (five genotypes, five trees each). The selected clones were among the most productive of the experiment and are hybrids from a diversity of species. **Table 1** shows the species, growth and wood properties obtained from disks at breast height of five elite *Eucalyptus* clones. The wood disks for this work were obtained at five different Heights (base, 25%, 50%, 75% and at 100% of the commercial height) for determining basic density and fiber characteristics. In addition, wood logs of approximately 50 cm were obtained at intermediate heights regarding those previously mentioned. These logs were chipped into small sized pieces (around 5 mm) and kept stored before kraft pulping. **Figure 2** illustrates the sampling process and respective analysis.

2.2 Fiber morphology and basic wood density

Blocks of approximately 1 cm^3 were sampled along the radius for fiber measurements from the pith to bark. Block samples were macerated with acetic acid and hydrogen peroxide, stained with Safranin and assembled in microscope slides. Images were obtained using a Zeiss Axio Scope A1 microscope and a digital camera (Canon A640). The length, diameter, wall thickness, and lumen width of fibers were

Clone	Genotype/hybrid	MAI	VA	SL	TL	HOL	EXT	ASH
		$m^3\ h^{-1}\ yr^{-1}$				%		
A	Spontaneous hybrid *E. urophylla*	50.9	11.8	3.5	27.5	66	6.3	0.25
B	*Eucalyptus pellita* x *E. grandis*	49.1	12.6	3.5	28.3	65.9	5.5	0.4
C	*Eucalyptus camaldulensis* x *Eucalyptus tereticornis*	34.6	14.4	3.4	28.4	66.2	5.1	0.34
D	*E. grandis* x (*E. urophylla* x *E. tereticornis*)	48.7	11.2	3.2	27.9	68	3.7	0.4
E	*E. grandis* x *E. urophylla*	68.3	12.2	3.6	28.6	66.5	4.6	0.35

MAIvol–mean annual volume increment; VA–vessels area; SL–soluble lignin; TL–total lignin; HOL–holocellulose; EXT–extractive contents.

Table 1.
Species, productivity, vessels area and chemical composition of wood of the five selected Eucalyptus *elite-clones.*

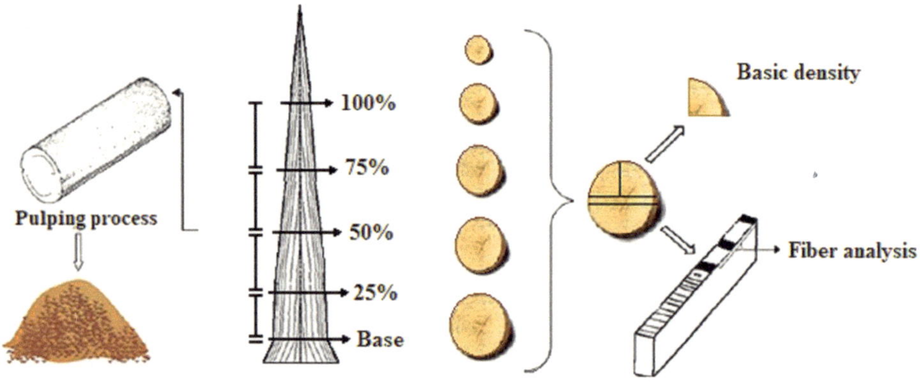

Figure 2.
Sampling process and respective analysis of the Eucalyptus clones wood.

measured using the Image Pro-Plus Software (version 5.0) program as recommended by International Association of Wood Anatomists (IAWA) [21].

Next, relationships were established between individual fiber dimensions and quality indices to evaluate the wood morphological properties for paper purposes, such as the Runkel ratio, wall proportion, flexibility coefficient, slenderness ratio and Luce's shape factor. These indices were analyzed according to the categories established by Barrichelo and Brito [22] and Foelkel et al. [23], and calculated according to the following equations:

$$Runkel\ ratio\,(RR) = 2w\,/\,d \tag{1}$$

$$Wall\ proportion\,(WP) = (2x\,/\,D)\times 100 \tag{2}$$

$$Flexibility\ coefficient\,(FC) = d\,/\,D \quad (3)$$

$$Slenderness\ ratio\,(SR) = L\,/\,D \quad (4)$$

$$Luce'shape\ factor\,(LSF) = \left(D^2 - d^2\right) / \left(D^2 + d^2\right) \quad (5)$$

In which: w is the cell wall thickness, D is the fiber diameter, d is the fiber lumen width, and L is the fiber length.

The basic wood density was determined according to ASTM D2395–17 using wood wedges (approximately 1/4) obtained from each disk.

2.3 Pulping process

All pulping processes were performed in triplicate in a rotating digester containing eight capsules with capacity of 10 L. An alkaline curve with four active alkali levels was performed under fixed kraft pulping conditions for each genotype (**Table 2**). The alkali levels were selected from previous tests in order to determine the dosage of active alkali (AA) required to obtain a kappa number 18.0 ± 0.5. The analysis of residual alkali in black liquor was performed according to SCAN-N 2:88 modified. Pulp yields and consumed alkali (difference between applied and residual alkali) were calculated, and the pulps' kappa number was determined according to TAPPI T 236 om-99. The other calculated parameters were wood specific consumption [24] and mean annual pulp increment [15], according to the following equations:

$$WSC = \frac{1}{BD \times PY} \times 0.9 \quad (6)$$

in which: WSC = wood specific consumption ($m^3\ t^{-1}$); BD = basic density ($g\ cm^{-3}$); PY = pulp yield (in decimal).

$$MAIpulp = \frac{MAIvol \times BD\ x\ PY}{1111} \quad (7)$$

in which: MAIpulp = mean annual increment of pulp ($t\ h^{-1}\ yr^{-1}$); MAIvol = mean annual increment of volume ($m^3\ h^{-1}\ yr^{-1}$); BD = basic density ($kg\ m^{-3}$); PY = pulp yield (%).

Parameter	Condition
Active alkali, % (NaOH based)	22, 24, 26 e 28
Sulphidity, %	25
Dry chip mass, g	70
Relation liquor/wood	04:01
Maximum temperature, °C	166
Time at maximum temperature, min	90
Heating time, min	60
H factor	780

Table 2.
Kraft pulping conditions.

2.4 Data analysis

The data were submitted to normality [25] and homogeneity tests of residual variances [26]. The effect of genetic variation between clones was evaluated for all the quantified wood variables by univariate analysis of variance (ANOVA). The averages were compared by the Tukey test (α = 0.05) when the effect of clones was significant, except the morphological ratios which were descriptively analyzed. The kraft pulping properties were evaluated using linear and polynomial regression models [27], where Y represents the kappa number, yield or consumed alkali, and Xi the alkali charge used. Pearson's correlations were used to examine expected relationships between wood properties. A principal component analysis (PCA) was also performed to recognize the most important parameters for the studied clones and their relationships. The analysis was performed using a correlation matrix with 95% statistical significance. The 95% ellipse limits were also used to test the variables in relation to the clones [28].

3. Results

3.1 Wood characteristics

The average values of fiber dimensions, morphological ratios and basic density are summarized in **Table 3**. Fiber morphology showed considerable differences between genotypes, except for length measurements. Fiber length corresponded from 813.6 μm (clone E) to 836.6 μm (clone D); fiber diameter from 16.9 μm (clone D) to 18.5 μm (clone B); fiber wall thickness from 3.8 μm (clone B) to 4.5 μm (clone D); and lumen width from 8.0 μm (clone D) to 10.9 μm (clone B).

Clone		**Wood**				**Fibers ratio**				**BD**
	FL	FD	FWT	FLW	RR	WP	FC	SR	LSF	
		(μm)				**(%)**				**(kg m^{-3})**
A	833.2^{a}	18.1^{a}	4.4a	9.4^{b}	1.05	50.2	0.50	45.9	0.58	489^{ab}
	(11.8)	(5.3)	(16.4)	(14.3)	(9.6)	(3.4)	(3.1)	(4.5)	(3.4)	(6.1)
B	834.4^{a}	18.5^{a}	3.8^{b}	10.9^{a}	0.76	43.5	0.57	45.5	0.49	446^{b}
	(8.2)	(11.2)	(16.9)	(10.0)	(8.7)	(5.0)	(3.5)	(5.9)	(5.5)	(5.1)
C	815.3^{a}	17.8^{ab}	4.1^{ab}	9.5^{b}	0.93	48.7	0.52	46.0	0.56	472^{ab}
	(8.1)	(5.1)	(13.6)	(10.7)	(15.4)	(9.0)	(7.9)	(3.0)	(9.7)	(3.3)
D	836.6^{a}	16.9^{b}	4.4^{a}	8.0^{c}	1.19	54.8	0.46	49.7	0.64	501^{a}
	(9.5)	(5.1)	(18.4)	(12.6)	(17.5)	(8.0)	(8.9)	(2.8)	(8.2)	(2.8)
E	813.6^{a}	18.2^{a}	4.2^{ab}	9.9^{b}	0.92	49.7	0.55	44.8	0.55	495^{a}
	(8.6)	(4.2)	(9.6)	(6.1)	(9.2)	(5.0)	(4.2)	(3.7)	(5.4)	(4.1)

FL–fiber length; FD–fiber diameter; FWT–fiber wall thickness; FLW–fiber lumen width; RR–Runkel ratio; WP–Wall proportion; FC–flexibility coefficient; SR–slenderness ratio; LSF–Luces's shape factor; BD–basic density. Means and coefficients of variation (%). Means followed by the same letter do not differ from each other by the Tukey test (p > 0.05).

Table 3.
Fiber dimensions, morphological ratios, and basic density of the five elite Eucalyptus spp. clones.

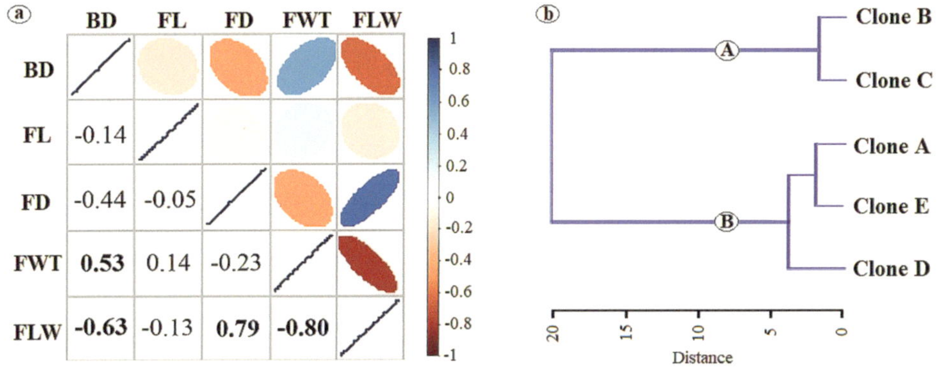

Figure 3.
Pearson's correlation (a) and hierarchical cluster analyses (b) for fiber dimensions and wood density of Eucalyptus *spp. clones. BD–basic density; FL–fiber length; FD–fiber diameter; FWT–fiber wall thickness; FLW–fiber lumen width.*

Regarding the morphological ratios, clone B had the lowest RR (Runkel ratio), WP (wall proportion) and LSF (Luce's factor): 0.76, 43.5%, and 0.49, respectively. The highest values for the same ratios were found for clone D (1.19 to RR, 54.8 to WP and 0.64 to LSF). In the same way but with opposite results, clone B showed the highest flexibility coefficient (0.59), and clone D presented the lowest value (0.46). Slenderness ratio values ranged from 44.8 (clone E) to 49.7 (clone D).

Mean wood density ranged from 446 to 501 kg m^{-3}, with the highest basic density for clones D and E and differing significantly between clone B. In **Figure 3a** it can be verified that wood density presented a significant inverse correlation with the fiber lumen width (p-value = 0.006) and a positive correlation with fiber wall thickness (p-value = 0.02). Correlations are also verified between fiber diameter and fiber wall thickness (p-value <0.001) and between fiber wall thickness and fiber lumen width (p-value <0.001). Taking to account the fiber dimensions and basic density, cluster analysis showed the formation of two main groups (**Figure 3b**). Group 1 comprised clones B and C, while group 2 contained clones A, D and E.

3.2 Kraft pulping properties

Figure 4 shows the graphs and equations established to mathematically determine the alkali charge to obtain the kappa number ±18. **Table 4** presents these corresponding values, as well as other technological parameters calculated for the target kappa. Clones decrease the kappa number at an average range rate of 8% (clone D) to 13% (clone A). In the pulp delignification degree for Kappa ±18, the wood from the five genotypes produced pulps with distinct characteristics, except for pulp yield which did not significantly differ between them with values ranging from 51.3–52.8%. Clone C used the lowest alkali to obtain a kappa number ± 18 (23.2%), significantly differing between clones A and B. Consumed alkali ranged between 32.0–40.6 g L^{-1}, with the lowest value for clone D, and with no significant difference between clone E. Clone C obtained the lowest residual effective alkali value (13.3 g L^{-1}), while clones B and D showed the highest values for this parameter (20.0 and 20.3 g L^{-1}, respectively). Clones A, D and E presented the lowest wood specific consumption, with values close to 3.5 m^3 t^{-1}, and clone E showed the best performance in terms of pulp productivity (MAIpulp = 19.8 t h^{-1} yr^{-1}).

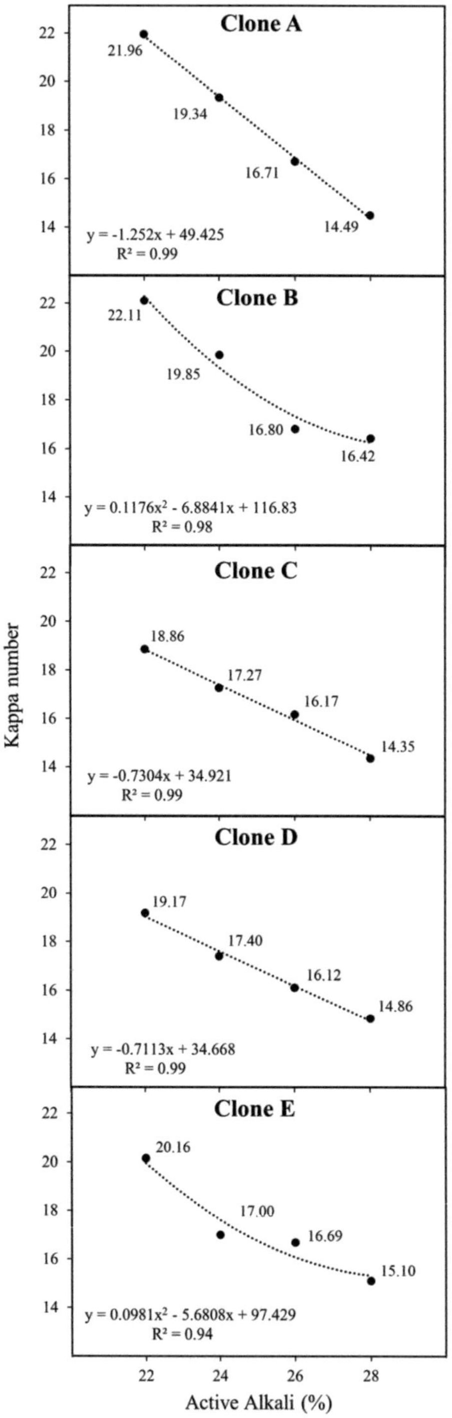

Figure 4.
Kappa number curves as a function of the active alkali applied to the five Eucalyptus *spp. clones. Adjusted-regression for higher correlation coefficient.*

Clone	Alkali charge	Yield	Consumed alkali	REA	WSC	MAIpulp
	%		g.L^{-1}		m^3.t^{-1}	t h^{-1} yr^{-1}
A	25.1^{ab}	52.3^a	40.6^a	16.0^b	3.5^c	14.7^b
B	25.5^a	51.3^a	38.5^a	20.0^a	3.9^a	12.5^d
C	23.2^c	52.8^a	39.7^a	13.3^c	3.6^b	9.5^e
D	23.5^{bc}	52.7^a	32.0^b	20.3^a	3.4^c	14.1^c
E	23.5^{bc}	52.2^a	35.3^{ab}	17.0^b	3.5^c	19.8^a

REA–residual effective alkali; WSC–wood specific consumption; MAIpulp–mean annual increment of pulp. Means followed by the same letter in the same column do not differ from each other by the Tukey test ($p > 0.05$).

Table 4.
Kraft pulping characterization for Kappa number ± 18 of the five Eucalyptus *spp. clones.*

3.3 Principal components analysis

Figure 5 shows the ordering of eigenvectors and the measures of similarity between *Eucalyptus* spp. clones. Clones were preserved in their characteristics but showed proximity between them. Regarding the study parameters, MAIpulp, consumed alkali, fiber lumen width and wood specific consumption are those which represented the largest variances in the analysis, determining the distribution of clones and variables. Three main groups of variables were formed: group 1 containing fiber diameter (FD), alkali charge (AC) and fiber lumen width (FLW), negatively related

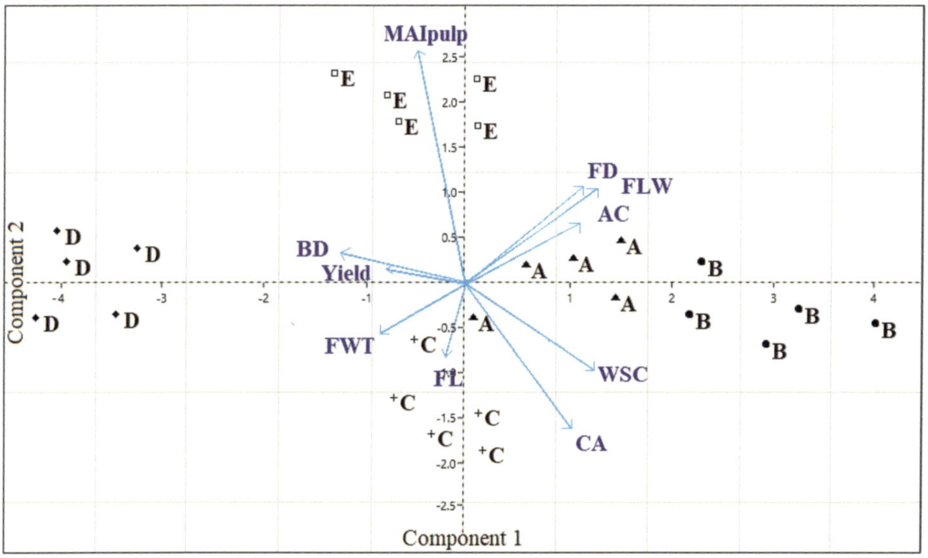

Figure 5.
Principal component analysis of the fiber dimensions, basic density and kraft pulping parameters of the five Eucalyptus *spp. clones. FL–fiber length; FD–fiber diameter; FWT–fiber wall thickness; FLW–fiber lumen width; BD- basic density; AC–Alkali charge; CA–consumed alkali; WSC–wood specific consumption and MAIpulp–mean annual pulp increment (significant variables according to Boot N x 10.000 criteria and their percentage of variance; MAIpulp –69.9%; CA – 45.2%; FLW- 42.7%; WSC – 37.5%).*

to clone D; group 2 formed by consumed alkali (CA) and specific wood consumption (WSC) directly related to clone B, and inversely to clone D; and group 3 formed by yield, fiber wall thickness (FWT) and basic density (BD) directly related to clones D, and negatively to clone B.

4. Discussion

4.1 Wood traits demonstrate significant variations among genotypes

Although there is a clear association between climate and wood properties, genetic factors influence their expression [5, 29] and shape the adaptive responses [30, 31]. This explains why the best performing *Eucalyptus* clones (elite clones) selected for this study are genetic materials which have relative drought tolerance [4, 31]. Thus, a selection metric of *Eucalyptus* clones based on traits for drought tolerance and high productivity combined with versatile wood properties for pulp and paper offers a good perspective for the current world pulp and paper industry.

Similar fiber dimensions were reported in the literature for young *Eucalyptus* trees with fiber length between 670 and 1040 μm, fiber diameter between 16 and 19 μm, fiber wall thickness between 3 and 5 μm and lumen width between 7 and 9.5 μm [16, 32, 33]. Wood-fiber characteristics are among the first indicators to be evaluated for screening potential fibrous raw materials for paper production. Fiber dimensions are generally related to collapsibility and flexibility properties, which are strongly associated with paper strength and surface properties [34, 35].

The lowest values presented by clone B for RR, WP and LSF is related to the fact that this genotype has comparatively thin walled fibers. It is assumed that an RR value less than 1.0 is favorable to produce an inter-fiber contact in manufacturing paper using hardwoods, and greater than 1.5 is not recommended. Percentages below 60% are indicated for the wall proportion, while Luce's factor is a ratio related to the final paper sheet density and can be used for specific selection [22, 23]. In this sense, all clones have adequate wall proportion, and clones B, C and E presented better RR values. According to [36], flexibility coefficient values between 0.50–0.75, as verified in this study (except for clone D), classify the fibers as flexible, which tend to form a highly resistant paper when intertwined. Finally, the slenderness ratio is a quality directly related to paper tear index, for which values above 50 are preferable. None of the clones showed a higher value as indicated, with clone D being the highest.

The values presented for basic wood density must be classified as average, meaning 450–540 kg m^{-3}, considered acceptable for commercial pulpwood and are similar to previous reports for young *Eucalyptus* trees between 450 and 500 kg m^{-3} [37, 38]. According to Gomide et al. [39], the projects to increase production capacity and to deploy new factories have prioritized the use of wood with a density close to 500 kg m^{-3}, as verified for the study clones, with an emphasis on clones D (*E. grandis* x *E. urophylla* x *Eucalyptus tereticornis*) and E (*E. grandis* x *E. urophylla*).

Significant correlations between fiber dimensions and density reflect the intrinsic. association between these properties. Barroto et al. [11] suggest a mediated effect of fiber wall fraction on the functionality of *Eucalyptus* trees through vessel dimensions and basic density. It is possible to note that clone D displayed higher fiber wall proportion and wood density, however it also had the lowest vessel area (see **Table 1**). Thus, a disadvantage is verified for this clone in terms of fibrous raw materials (quality indices), although the higher average density results in lower wood specific consumption. Similar behavior can be observed for clone A (*E. urophylla*).

4.2 The evaluated clones show potential for cellulose pulp in central-western Brazil

Despite the large variation observed in wood properties, it is possible to affirm that all *Eucalyptus* clones presented quality as raw material to make paper and good pulping performance. The alkali charge values found in the literature for approximate target kappa are lower than those presented in this study [40, 41]. However, Gomide et al. [42] cites alkali percentages between 20.1 and 23.7% for kappa ±18 in assessing 75 wood samples of *Eucalyptus* clones from different Brazilian regions, thus evidencing the great variability existing for this technological parameter depending on the genotype and the environmental influences on growth. These high alkali charges required can be due to relatively high extractive content presented by genotypes (see **Table 1**), which increases the active alkali demand for pulping the wood to a given kappa number [43, 44].

Increasing pulp yield is a major goal of a chemical process, being related to the reduction of residual effective alkali (REA) at the end of cooking [45]. In this sense, higher value of consumed alkali represents better kraft pulping performance. Although the five *Eucalyptus* clones showed similar yield values, the same was not observed for the consumed alkali and residual effective alkali rates. According to Ribeiro et al. [45], the REA varies between 4 and 18 g L^{-1} considering different mills, while the range of 7–9 g L^{-1} is more common. Given this, the result is that the clones B and D reached high values for this parameter. Segura et al. [46] presented consumed alkali values on different levels of kappa number around 35–45 g L^{-1} for *E. grandis* x *E. urophylla* at 6 years old, thus corroborating the results presented herein. For *Eucalyptus* trees grown in southeastern Brazil, Gouvêa et al. [40] mentions pulp yield values from 50.3–52.9%, and Ferreira et al. [47] relates pulp yield values between 50 and 55% for *Eucalyptus* commercial clones at seven-nine years old. Similar results were reported by other authors [48].

The pulping yield directly influences the wood specific consumption - WSC, which in this work, was obtained taking into account only the yield in each pulping, disregarding the losses in the subsequent processes. Segura et al. [46] emphasizes that the basic density, which is an associated key property to wood specific consumption, has an important role in pulping costs and the densest materials present lowest wood specific consumption, reducing the wood and process costs. At the same time, higher densities may imply in low permeability of the wood by the liquor, thereby requiring high chemical usage and energy [49]. This is why density values within a medium range which configure a high proportion of wood in balance with good efficiency in the pulping process is given importance.

The best performance of clone E (*E. grandis* x *E. urophylla*) in terms of mean annual increment of pulp is due the high wood productivity (see **Table 1**, mean annual volume increment), which emphasizes the relevance of this attribute in assessing potential on a large scale. Clone C (*Eucalyptus camaldulensis* x *E. tereticornis*) presented the greatest disadvantage for this parameter. Clone B (*E. grandis* x *Eucalyptus pellita*) presented satisfactory growth in contrast with relatively inferior pulping performance and displayed less dense wood associated to wider fibers and lumen.

As can be seen, variations in the wood properties act dynamically on raw material quality and technological performance of *Eucalyptus* spp. clones. The magnitude of changes in the wood properties in response to the site did not evidence negative impacts on the pulping process efficiency of *Eucalyptus* clones at 4 years in comparison with the literature, except for alkaline charge. With respect to large-scale pulp production, there is a strong influence of growth potential and adaptive capacity of different hybrids. In this sense, the *E. grandis* x *E. urophylla* hybrid having highest wood volume stands out. Finally, the genetic material choice determined by factors which are related to environmental conditions, together with an assessment on the particular parameters related to end-use, as guided by the present study, plays a very important role in efficient breeding selection. However, it is important to emphasize that other traits should be explored in considering the overall assessment of the kraft pulp potential of a wood, such as the cellulose and lignin composition and fiber morphological and handsheet properties.

5. Conclusion

The five elite *Eucalyptus* clones at 4 years generally presented wood quality for pulping or paper end-use with adequate values for most fiber ratios and average basic density. Good technological performance was obtained with respect to pulp yield. In contrast, genotypes presented a high required alkali charge to obtain a kappa number ±18.0 as a disadvantage. Overall, clone E (*E. grandis* x *E. urophylla*) stands out among the analyzed genotypes in terms of combined characteristics. Clone D (*E. grandis* x *E. urophylla* x *E. tereticornis*) presented inferior fiber quality as a consequence of thicker cell wall associated to thinner fibers. Conversely, clone B (*E. pellita* x *E. grandis*), which has the lowest wood density associated to wider lumen and fibers, displayed relatively lower pulping performance.

The prior selection of *Eucalyptus* genotypes with drought tolerance and high productivity traits, associated to the significant variations found among them in this present work for important parameters associated with wood properties and pulpability, represents the basis for an efficient evaluation based on pulp potential in aiming to select genotypes for commercial plantations. Future studies to explore other important quality traits must be performed.

Acknowledgements

The authors acknowledge the Chemistry, Pulp and Energy Laboratory and the Wood Anatomy and Tree-Ring Laboratory of São Paulo University (USP) and the Wood Quality and Bioenergy Laboratory of the Federal University of Goiás.

Conflict of interest

The authors declare no conflict of interest.

Author details

Deborah Rodrigues de Souza Santos[1*], Camila Sarto[1], Rafael Fernandes dos Santos[2], Júlia Lôbo Ribeiro Anciotti Gil[1], Carlos de Melo e Silva-Neto[3], Regina Maria Gomes[1], Evandro Novaes[4], Carlos Roberto Sette-Junior[2], Mario Tomazello-Filho[1], Rafael Tassinari Resende[2] and Matheus Peres Chagas[2]

1 University of São Paulo, Piracicaba, Brazil

2 Federal University of Goiás, Goiânia, Brazil

3 Federal Institute of Goiás, Goiânia, Brazil

4 Federal University of Lavras, Lavras, Brazil

*Address all correspondence to: deborah.rodrigues@usp.br

References

[1] IBA. Annual Report 2020. 2020. Available from: https://iba.org/datafiles/publicacoes/relatorios/relatorio-iba-2020

[2] Buol SW. Soils and agriculture in central-west and north Brazil. Science in Agriculture. 2009;**66**:697-707. DOI: 10.1590/s0103-90162009000500016

[3] Martins GS, Freitas NC, Máximo WPF, Paiva LV. Gene expression in two contrasting hybrid clones of *Eucalyptus camaldulensis* x *Eucalyptus urophylla* grown under water deficit conditions. Journal of Plant Physiology. 2018;**229**:122-131. DOI: 10.1016/j.jplph.2018.07.007

[4] Gonçalves JLM, Alvares CA, Rocha JHT, Brandani CB, Hakamada R. Eucalypt plantation management in regions with water stress. Southern Forests. 2017;**79**:169-183

[5] Raymond CA. Genetics of Eucalyptus wood properties. Annals of Forest Science. 2002;**59**:525-531. DOI: 10.2989/20702620.2016.1255415

[6] Gomes FJB, Colodette JL, Burnet A, Batalha LAR, Santos FA, Demuner IF. Thorough characterization of Brazilian new generation of Eucalypt clones and grass for pulp production. International Journal for Research. 2015;**2015**:571-588. DOI: 10.1155/2015/814071

[7] Trugilho PF, Goulart SL, De Assis CO, Couto FBS, Alves ICN, Protásio TDP, et al. Growth characteristics, chemical, physical composition and dry mass estimation of wood in clones and young *Eucalyptus* species. Ciencia Rural. 2015;**45**:661-666. DOI: 10.1590/0103-8478cr20130625

[8] de Almeida MNF, de Vidaurre GB, Pezzopane JEM, Lousada JLPC, Silva MECM, Câmara AP, et al. Heartwood variation of *Eucalyptus urophylla* is influenced by climatic conditions. Forest Ecology and Management. 2020;**458**:1-10. DOI: 10.1016/j.foreco.2019.117743

[9] Barbosa TL, Oliveira JT d S, Rocha SMG, Câmara AP, Vidaurre GB, Rosado AM, et al. Influence of site in the wood quality of *Eucalyptus* in plantations in Brazil. Southern Forests. 2019;**81**:247-253. DOI: 10.2989/20702620.2019.1570453

[10] Rocha SMG, Vidaurre GB, PezzopaneJEM, AlmeidaMNF, CarneiroRL, Campoe OC, et al. Influence of climatic variations on production, biomass and density of wood in *Eucalyptus* clones of different species. Forest Ecology and Management. 2020;**473**:118290. DOI: 10.1016/j.foreco.2020.118290

[11] Barotto AJ, Monteoliva S, Gyenge J, Martinez-Meier A, Fernandez ME. Functional relationships between wood structure and vulnerability to xylem cavitation in races of *Eucalyptus globulus* differing in wood density. Tree Physiology. 2018;**38**:243-251. DOI: 10.2989/20702620.2019.1570453

[12] Pfautsch S, Harbusch M, Wesolowski A, Smith R, Macfarlane C, Tjoelker MG, et al. Climate determines vascular traits in the ecologically diverse genus *Eucalyptus*. Ecology Letters. 2016;**19**:240-248. DOI: 10.1111/ele.12559

[13] Hacke UG, Sperry JS, Pockman WT, Davis SD, McCulloh KA. Trends in wood density and structure are linked to prevention of xylem implosion by negative pressure. Oecologia. 2001;**126**:457-461. DOI: 10.1007/s004420100628

[14] Jacobsen AL, Ewers FW, Pratt RB, Paddock WA, Davis SD. Do xylem fibers affect vessel cavitation resistance? Plant Physiology. 2005;**139**:546-556. DOI: 10.1104/pp.104.058404

[15] Gallo R, Pantuza IB, dos Santos GA, de Resende MDV, Xavier A, Simiqueli GF, et al. Growth and wood quality traits in the genetic selection of potential *Eucalyptus dunnii* Maiden clones for pulp production. Industrial Crops and Products. 2018;**123**:434-441. DOI: 10.1016/j.indcrop.2018.07.016. DOI: 10.1016/j.indcrop.2018.07.016

[16] Pirralho M, Flores D, Sousa VB, Quilhó T, Knapic S, Pereira H. Evaluation on paper making potential of nine *Eucalyptus* species based on wood anatomical features. Industrial Crops and Products. 2014;**54**:327-334. DOI: 10.1016/j.indcrop.2014.01.040

[17] Wimmer R, Downes G, Evans R, French J. Effects of site on fibre, kraft pulp and handsheet properties of *Eucalyptus globulus*. Annals of Forest Science. 2008;**65**:602-602. DOI: 10.1051/forest:2008039

[18] Freitas TP, da S Oliveira JT, Vidaurre GB, Rodrigues BP, Freitas TP, da S Oliveira JT, et al. Environmental effect on chemical composition of *Eucalyptus* clones wood for pulp production. Cerne. 2018;**24**:219-224. DOI: 10.1590/01047760201824032558

[19] Reis CF, Moraes AC, Pereira AV, Aguiar AV, Sousa VA, Borges HMD. Diagnosis of the planted forest sector in the State of Goiás. Brasília- DF: Embrapa Forestry; 2015. p. 140

[20] Cardoso MRD, Marcuzzo FFN, Barros JR. Climatic classification of Köppen-Geiger for the state of Goiás and federal district. Acta Geográfica. 2014;**8**:40-55. DOI: 10.5654/actageo2014.0004.0016

[21] IAWA. List of microscopic features for hardwood identification. International Association of Wood Anatomists Bulletin New Series. 1989;**10**:219-332

[22] Barrichelo LEG, Brito JO. Wood of *Eucalyptus* species as raw material for the pulp and paper industry. Brasília: PRODEPEF-13; 1976. p. 145

[23] Foelkel CEB, Zvinakevicius C, Andrade J, Medeiros SJ. Tropical *Eucalyptus* in Cellulose Production. Belo Oriente: Cenibra; 1978. p. 31

[24] Segura TES, Dos Santos JRS, Sarto C, Jr DS, FG. Effect of kappa number variation on modified pulping of *Eucalyptus*. BioResources. 2017;**11**:9842-9855. DOI: 10.15376/biores.11.4.9842-9855

[25] Shapiro SS, Wilk MB. An analysis of variance test for normality complete samples. Biometrika. 1965;**52**:591-611. DOI: 10.2307/2333709

[26] Levene H. Robust testes for equality of variances. In: Contributions to Probability and Statistics. 1st ed. Palo Alto: Stanford University; 1960. pp. 278-292

[27] Gomide JL, Fantuzzi Neto H, Leite HG. A laboratory technique to establish *Eucalyptus* sp. wood quality for kraft pulp production. Revista Árvore. 2004;**28**:443-450. DOI: 10.1590/S0100-67622004000300015

[28] Couto AM, de Protásio TP, Trugilho PF, Neves1 TA, de Sá VA. Multivariate analysis to evaluation of *Eucalyptus* clones for bioenergy production. Cerne. 2013;**19**:525-533. DOI: 10.1590/S0104-77602013000400001

[29] de Costa SEL, do Santos RC, Vidaurre GB, Castro RVO, Rocha SMG,

Carneiro RL, et al. The effects of contrasting environments on the basic density and mean annual increment of wood from *Eucalyptus* clones. Forest Ecology and Management. 2020;**458**:1-10. DOI: 10.1016/j.foreco.2019.117807

[30] Bourne AE, Creek D, Peters JMR, Ellsworth DS, Choat B. Species climate range influences hydraulic and stomatal traits in *Eucalyptus* species. Annals of Botany. 2017;**120**:123-133. DOI: 10.1093/aob/mcx020

[31] Saadaoui E, Ben Yahia K, Dhahri S, Ben Jamaa ML, Khouja ML. An overview of adaptative responses to drought stress in *Eucalyptus* spp. Forestry Studies. 2017;**67**:86-96. DOI: 10.1515/fsmu-2017-0014

[32] Baldin T, Marchiori JNC, Nisgoski S, Talgatti M, Denardi L. Anatomy of wood and pulp and paper production potential of four young species of *Eucalyptus* L'Hér. Ciência da Madeira. 2017;**8**:114-126. DOI: 10.12953/2177-6830/rcm.v8n2p114-126

[33] Rao RV, Shashikala S, Sreevani P, Kothiyal V, Sarma CR, Lal P. Within tree variation in anatomical properties of some clones of *Eucalyptus tereticornis* Sm. Wood Science and Technology. 2002;**36**:271-285. DOI: 10.1007/s00226-002-0139-3

[34] Bronkhorst CA. Modelling paper as a two-dimensional elastic - plastic stochastic network. International Journal of Solids and Structures. 2003;**40**:5441-5454. DOI: 10.1016/S0020-7683(03)00281-6

[35] Clark JA. Pulp Technology and Treatment for Paper. São Francisco: Miller Freeman Publications; 1978. p. 752

[36] Bektas I, Tutus A, Eroglu H. A study of the suitability of Calabrian Pine (*Pinus burtia*, Jen) for pulp and paper manufacture. Turkish Journal of Agriculture and Forestry. 1999;**23**:589-597. DOI: 10.3906/tar-7-97115

[37] Castro AFNM, Castro RVO, de Carneiro ACO, dos Santos RC, Carvalho AMML, Trigilho PF, et al. Correlations between age, wood quality and charcoal quality of *Eucalyptus* clones. Revista Árvore. 2016;**40**:551-560. DOI: 10.1590/0100-67622016000300019

[38] Hsing TY, de Paula NF, de Paula RC. Dendrometric, chemical characteristics and basic density of wood from *Eucalyptus grandis* x *Eucalyptus urophylla* hybrids. Cienc Florest. 2016;**26**:273-283. DOI: 10.5902/1980509821119

[39] Gomide JL, Colodette JL, Chaves de Oliveira R, Silva CM. Technological characterization of the new generation of *Eucalyptus* clones in Brazil for kraft pulp production. Revista Árvore. 2005;**29**:129-137. DOI: 10.1590/S0100-67622005000100014

[40] de Gouvêa AFG, Trugilo PF, Colodette JL, Lima JT, da Silva JRM, Gomide JL. Wood evaluation and kraft pulping in eucalypts clones. Revista Árvore. 2009;**33**:1175-1185. DOI: 10.1590/S0100-67622009000600020

[41] Queiroz SCS, Gomide JL, Colodette JL, Oliveira RC. Effect of wood basic density on Kraft pulp quality of hybrid *Eucalyptus grandis* W. Hill ex Maiden X *Eucalyptus urophylla* S.T. Blake clones. Revista Árvore. 2004;**28**:901-909. DOI: 10.1590/S0100-67622004000600016

[42] Gomide JL, Fantuzzi Neto H, Regazzi AJ. Analysis of wood quality criteria of *Eucalyptus* wood for kraft pulp production. Revista Árvore. 2010;**34**:339-344. DOI: 10.1590/S0100-67622010000200017

[43] da Magaton AS, Colodette JL, de Gouvêa AFG, Gomide JL, Muguet MCDS, Pedrazzi C. *Eucalyptus* wood quality and its impact on Kraft pulp production and use. Tappi Journal. 2009;**8**:32-39

[44] Sansígolo CA, da Ramos ÉS. Quality of wood and pulp from a clone of *Eucalyptus grandis* planted at three locations. Cerne. 2011;**17**(47-60). DOI: 10.1590/S0104-77602011000100006

[45] Ribeiro RA, Colodette JL, Vaz Júnior S. Effect of residual effective alkali on *Eucalyptus* kraft pulp yield and chemistry. Cerne. 2018;**24**:408-419. DOI: 10.1590/01047760201824042593

[46] Segura TES, Zanão M, Santos JRS, Silva FG. Kraft pulping of the main hardwoods used around the world for pulp and paper production. In: TAPPI PEERS Conf Build a Sustain Futur. Savannah. Georgia; 2012. pp. 593-600

[47] Ferreira CR, Fantini Junior M, Colodette JL, Gomide JL, Carvalho AMML. Technological assessment of *Eucalyptus* wood clones: Part 1 - Wood quality for kraft pulp production. Science Forest. 2006;**70**:161-170

[48] de Morais PHD, Júnior DL, Colodette JL, da Morais EHC, Jardim CM. Influence of clone harvesting age of *Eucalyptus grandis* and hybrids of *Eucalyptus grandis* x *Eucalyptus urophylla* in the wood chemical composition an in kraft pulpability. Cienc Florest. 2017;**27**:237-248. DOI: 10.5902/1980509826462

[49] Torgovnikov G, Vinden P. High-intensity microwave wood modification for increasing permeability. Forest Products Journal. 2009;**59**:84-92